contents

19600

KV-296-626

thematic contents

Trade

geothermal power (17, 43, 39), sediments (25), sandstones (28), limestone (29, 37, 38), slate (32), millstone grit (37), coal (38), minerals (43, 44).

Hazards

earthquakes (10, 13, 24, 44–48), rift valleys (15, 16), Iceland (16), volcanoes (17, 23, 30, 40, 44), subduction (23), rubbish (23), orogeny (37), faulting (37), flash floods (38), climatic change (40), tsunami (44).

Green

fossil distribution (6), palaeomagnetism (8), theory of plate tectonics (10), rock cycles (11, 26, 27, 32), formation of earth (13, 19), isostasy (14, 20), life (15), sea floor spreading (16, 17, 22), cycle of erosion (21), plate margins (22), mountain building (23, 24), sea level changes (26–27), Milankovitch cycles (42).

Political

earthquake in Japan (23), reservoir rocks (29), mineral deposits (43), energy (43), hazards and risks (44), population density (44), planning for earthquakes (47).

Skills

judging ice movements (6), superposition (12), sea level changes (20), mapping features (23), sketching sections (30, 39), DME (45).

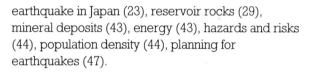

the geography collection is a series of texts designed for students of the new A-level geography syllabuses and has been written by a team of eminent geography teachers. The series provides an enlightened approach to the discipline, opening up the world of the geographer through contemporary text, clarifying illustrations, integrated questions and decision making exercises. The core text, *World Wide*, and Option Books contain guidelines for geographical skills and project work together with a glossary of important terms. A feature of the series are the themes that run through each book, signalled by icons, giving extra flexibility in the way that students can use the texts.

SECTION 1

19600

The geological background

KEY IDEAS

- **Some continents appear to fit together like a jigsaw.**
- **Early continental drift theories seemed convincing but were only circumstantial.**
- **Greater certainty required new evidence and mechanisms.**

JOINING CONTINENTS?

After the early European explorers of the sixteenth century returned home and produced their maps of South America and Africa, Francis Bacon and others observed that the bulge of Brazil would fit quite neatly into the shape of West Africa (Figure 1.1); and more recently it has been shown that an even better fit is achieved if the edges of the two **continental shelves**, rather than the relatively temporary present coastlines, are matched up instead. In the nineteenth century, in particular, other ideas and observations

gave support to the view that some continents were once joined, and in 1915 a German scientist and explorer, Alfred Wegener, brought the ideas together in his book *The Origin of the Continents and Oceans.* This produced convincing if circumstantial evidence that the continents had once been joined but had since drifted apart. However, as so often happens, Wegener's ideas were too far ahead of his time, and although there was plenty of discussion of the ideas of **continental drift**, the lack of a convincing mechanism prevented their general acceptance.

The evidence

Four types of evidence were proposed in support of continental drift.

1 The continental jigsaw
Apart from noting the obvious fit of Africa and South America, Wegener showed that mountain ranges in both hemispheres could be matched if the continents were brought together. A more recent proposal is shown in Figure 1.2 on page 6.

2 Past climates
Evidence of the deposits of ancient glaciers can be found in all the southern continents, even in areas which now experience equatorial and tropical climates. Furthermore, the indicators of the directions of ice movement make no sense unless the continents are assembled so that ice is shown to radiate from an

Figure 1.1 *The Atlantic jigsaw*

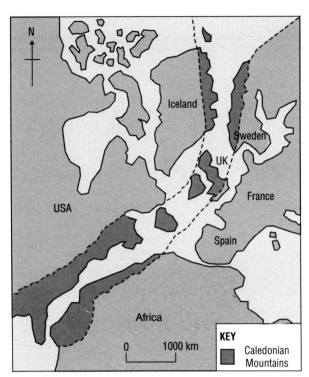

Figure 1.2 *A reconstruction of the position of the Caledonian Mountains (when formed about 400 million years ago) on either side of the North Atlantic*

Q1. *What features can indicate the direction of ice movement?*

Q2. *Figure 1.3a shows indicators of ice movement found in the southern continents. Use tracing paper to show how well they explain the pattern shown on Figure 1.3b.*

3 The distribution of fossils

The distribution of various plants, animals and fresh water fishes show that continents now separated by oceans must have been linked at one time. For example a 1 m long fossil reptile *Mesosaurus*, found only in fresh water sediments, has a limited distribution in south west Africa and Brazil; it is unlikely that it could have swum the present Atlantic, and so indicates that these continents were once joined. Similar species of fossil land plants are found on both sides of the South Atlantic, as are frogs, tortoises and fresh water fish which seem to have been able to move between the two until about 100 million years ago. Long 'land bridges' could be invoked to explain these distributions, but many of these seem very unconvincing when the distances involved and the depth of the intervening oceans are taken into account. Joining the continents together would produce a more straightforward explanation.

4 Crustal differences

By Wegener's time it had been realised that the material of the continents was lighter than that of the oceans and must be, in some way, floating on denser material beneath (Figure 1.4). The lighter **continental**

ice cap over central Africa. The single ancient continent which this seems to indicate is known by geologists as **Gondwanaland**. In a similar way, evidence of tropical conditions (such as coal deposits, evaporite deposits and sand dunes) can be matched if continents are fitted together.

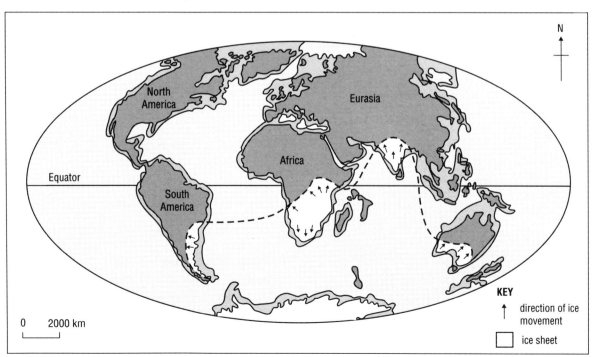

Figure 1.3a *Ice movement in Gondwanaland*

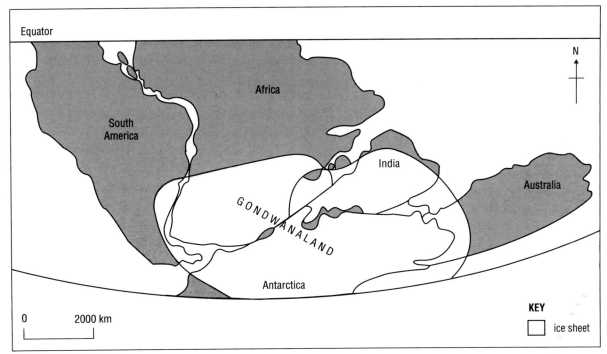

Figure 1.3b *Ice movement in Gondwanaland continued*

crust is often known as **sial** (silica + aluminium) and the denser material beneath as **sima** (silica + magnesium). Wegener argued that if the crust could move in a vertical direction, it might also move in a horizontal direction *'provided only that there are forces in existence which tend to displace continents, and that these forces last for geological epochs'*.

The nature of these forces was unknown at the time, and is still not fully understood today. Wegener however was convinced that there was enough empirical evidence to support his theory, but it was almost 50 years before it became generally accepted. The main problem was to explain how the sial continents could possibly 'plough their way' through the denser sima beneath. The answer when it came was one of classic simplicity: they don't!

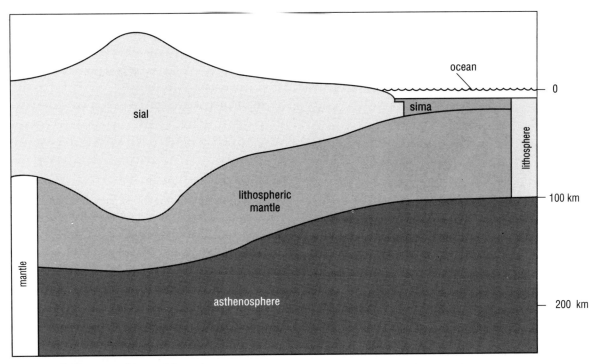

Figure 1.4 *The parts of the earth's crust*

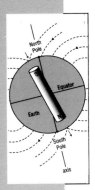

The new discoveries

- **Increasing knowledge confirmed that continents have drifted.**
- **By the late 1960s plate tectonics theory was generally accepted.**
- **Mechanisms have been suggested but little is known about the underlying causes.**

STUDIES OF PALAEOMAGNETISM

It may still be true that we know more about remote objects in the solar system than we do about the interior of our own planet. This was certainly true in the early 1960s when the first parts of the plate tectonics theory were starting to fall into place.

The earth is rather like a large bar magnet with the magnetic field at present lying at 11 degrees from the earth's rotational axis (Figure 2.1). Note that the angle of inclination combined with direction can indicate one's position on the globe better than direction alone. Although the magnetic poles change position to some extent, they appear to have remained reasonably close to their present positions over long periods.

Almost all rocks contain small quantities of iron compounds which, when solidified, become magnetised with respect to the magnetic pole. This is true for iron particles in igneous rocks, especially basalt lavas, and also, to some extent in sedimentary and metamorphic rocks. Thus, if a sample of rock of known age is taken, and the **palaeomagnetism** of its iron compounds is recorded, these will indicate its position with respect to the magnetic pole at the time of its formation. These studies of palaeomagnetism have made two different contributions to the understanding of how the continents move.

● For rocks formed recently, the palaeomagnets point to positions within the Arctic area; but the older the rocks are, the more they tend to point to other positions of the magnetic pole. Now if rocks of the same age from different continents all pointed the same way one might conclude that the magnetic pole had shifted. However this is not the case, and so it became apparent that the pole has stayed in the same place, but the continents have moved (Figure 2.2). It is as if a ship steering a crooked course recorded magnetic bearings every hour. One could plot its course by working

Figure 2.1 *The earth's magnetic field and angles of inclination (N.B. there is no bar magnet as such)*

backwards from its latest position. Thus the term for this information 'polar wandering' is misleading as it is the continents and not the magnetic pole which have wandered.

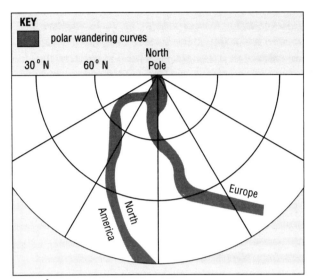

Figure 2.2 *Polar wandering curves*

When the curves for the last 500 million years are brought together, the two continents are found to have been joined until the curves separated as the Atlantic started to open some 200 million years ago.

● By 1960 research ships had towed magnetometers over much of the ocean floor and this resulted in a second surprise: the discovery that the earth's magnetic field has often 'flipped' so that the north magnetic pole switches at irregular intervals to being at the Antarctic. The reasons for this are not fully understood but a time chart of these **magnetic reversals** in the last 4.5

million years shows times when the magnetism was 'normal' and when it was 'reversed' (Figure 2.3). When rocks from the ocean floors were mapped to show normal and reversed magnetism it was found that they produced symmetrical patterns on either side of the lines of mountains in the oceans: the **mid-ocean ridges**.

Deep sea drilling

Since 1968 thousands of cores of rock have been drilled from the beds of the deep oceans by specially equipped vessels (Figure 2.4), the most famous being the Glomar Challenger. Using **radio-metric dating** techniques, it has been established that rocks get steadily older away from the mid-ocean ridges. In fact the patterns are again symmetrical and correspond closely with those of the magnetic reversals.

Figure 2.4 *Deep sea drilling vessel*

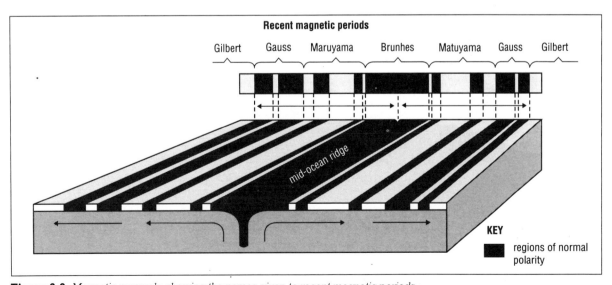

Figure 2.3 *Magnetic reversals, showing the names given to recent magnetic periods*

Earthquake zones

Not only do earthquakes occur in clear belts on the world map, but when deep and shallow earthquakes are plotted separately it is found that most of the deep ones are found alongside ocean trenches, with the deepest of all being the furthest away (Figure 2.5). Thus a section would show a zone of earthquakes (the **Benioff zone**) sloping away from the trench. The whole area where the plate is moving down into the earth is called the **subduction zone**.

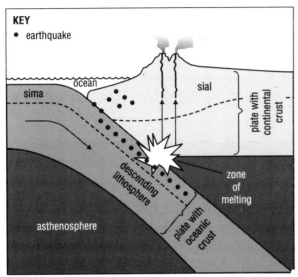

Figure 2.5 *Subduction of an ocean plate such as along the west coast of South America*

> **Q3.** *Make a sketch map from Figure 2.5. Show the deep trench, the deep and shallow earthquakes and the volcanoes.*
>
> **Q4.** *Another discovery was that the oldest ocean crust is only about 200 million years old, even though the oldest known continental crust, in Greenland, is over 450 000 million years old. Explain the significance of this discovery.*

The theory of plate tectonics

The theories of continental drift of Wegener and his followers had lacked a convincing mechanism. In 1930, Arthur Holmes proposed that convection currents within the earth might move the continents around like scum on the surface of a pan of boiling jam. However it was still very hard to see how the lighter sial continents could plough their way through the denser material of the sima crust.

By the 1960s it was realised that the crust is actually divided into a dozen or so large plates, plus a few smaller ones (see Figure 2.7, page 2). The edges of these can be mapped by plotting volcanic activity and earthquakes (Figure 2.6), and the plates themselves slide around the surface of the earth. In 1960 Professor Harry Hess of Princeton University postulated that the plates are being formed at the mid-ocean ridges, and move away from them sideways (**sea floor spreading**) to be consumed back into the earth at subduction zones where they meet (Figure 2.7). The key difference from earlier theories is that the continents are not ploughing through the sima but are themselves drifting as integral parts of the crustal plates.

Why do the plates move?

The precise reasons for all this movement are still uncertain but they are probably linked to convection currents within the earth (Figure 2.7). A major contributing factor is probably the weight of the material descending into the earth at subduction zones, as this probably drags the rest of the plates down behind.

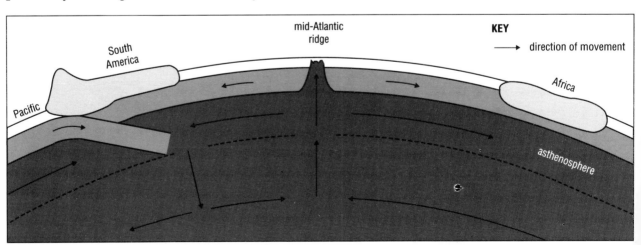

Figure 2.6 *Volcanic activity and earthquakes*

More geological background

KEY IDEAS

- **How are rocks dated?**
- **How do we know what the earth's interior is like?**

ROCKS

The rock cycle

All the rocks which first formed the crust of the earth were **igneous rocks**, formed from solidified molten rock (**magma**). These were then weathered and eroded and the resulting particles transported and deposited to form the **sedimentary rocks**. Sometimes, during earth movements, these two sorts of rock were subjected to tremendous heat and/or pressure which changed their composition and formed the **metamorphic rocks**. In extreme cases the rocks melt back into the magma and so a **rock cycle** develops (Figure 3.1). All these processes are still at work, producing new rocks all the time.

Figure 3.1 *The rock cycle*

Solid rock?

How can apparently solid rocks flow and bend in ways which may seem improbable? Easily! Think of crushing a drink can, of making a horse-shoe or a glass bottle, and of the way that glass in old windows is thicker at the bottom. All these 'solid' materials can bend given enough pressure (the drink can), heat (the horse-shoe) and time (the window). These causes of plastic deformation are enough to explain how rocks can flow and bend given the immense amounts of heat, pressure and time available for geological processes.

Dating the rocks

When reading British history one can understand many events so long as the order in which they occurred and what was happening in other countries at the same time is known. It makes understanding even better if one knows the actual dates concerned. In the same way, geologists deal with geological periods, first by getting them in the right order (this is known as **relative dating**) and, more recently, by discovering the dates in years (**absolute dating**).

Relative dating

This uses the field relationships of different rocks (Figure 3.2) to place them in the correct chronological order, and this is helped by the use of fossils which will also allow rocks to be correlated world-wide.

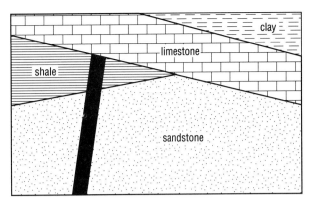

Figure 3.2 *A section through an imaginary cliff face. The thick black line represents an intrusion of igneous rock*

> **Q5.** *Study Figure 3.2 and list the eight geological events which produced this sequence, with the youngest (most recent) last.*

Absolute dating

Absolute dating (giving the dates in years) of rocks relies on the breakdown of certain radioactive minerals which decay at known rates. The result of the breakdown of a radioactive parent isotope is a stable daughter isotope which does not decay. For example, minerals in some lavas contain a parent isotope, Potassium 40. It takes 11 900 million years (its 'half life') for half of this to breakdown to its daughter atom, Argon 40. In another 11 900 million years another half has gone so that only a quarter of the parent atoms remain. As these quantities can now be measured, the difference between the proportions of the potassium and the argon can be calculated, and knowing the half life of potassium an absolute age can be obtained for the ancient igneous rock. This is similar to the radio-carbon dating techniques used by archaeologists and geomorphologists, but because the breakdown of the radioactive carbon is so fast (half life = 5730 years) it is of little value beyond 60 000 years ago.

> **Q6.** *In a granite a pair of minerals is found where a quarter are the parent atoms and three-quarters are the daughter atoms. The half life of this mineral is 710 million years. How long ago did the magma solidify?*
>
> **Q7.** *Suggest why such results would be less useful for dating sedimentary and metamorphic rocks.*

GEOLOGICAL TIME

Current estimates give the age of the solar system as about 5000 million years and the formation of the planet earth at about 4700 million years ago. Relatively little is known about early events but as we come closer to the present day our knowledge increases so that it is possible to divide time into increasingly short periods. Figure 3.3 gives a summary of the main sub-divisions commonly used, and shows a few of the main events. The major division at the base of the Cambrian represents a time when large numbers of fossils first appeared and this enabled the early geologists to construct the phanerozoic part of the column (i.e. since the Cambrian period) from fossil evidence and other relative dating methods.

> **Q8.** *Use library or other material to find out the main events in the development of life on earth. Include the dates of the first signs of life; the development of multi-cellular organisms; the first fish, reptiles and mammals; the first terrestrial plants and the first flowering plants; the main periods of mass extinction; and of people.*

The geological column

This column (Figure 3.3) is the basis of the study of the history of rocks and is needed in the same way as an historian needs a chronological chart.

ERA	PERIOD	START (Ma)	OROGENIES
Cainozoic	Quaternary Tertiary	0.01 65	←Alpine
Mesozoic	Cretaceous Jurassic Triassic	132 208 245	
Upper Palaeozoic	Permian Carboniferous Devonian	290 363 409	←Hercynian
Lower Palaeozoic	Silurian Ordovician Cambrian	439 510 570	←Caledonian
Pre-Cambrian		4700	Various

N.B: **Ma** = millions of years BP (before present).
Orogenies are periods of mountain building.

Figure 3.3 *The geological column*

The development of life is probably related to the gradual build up of free oxygen to provide for the respiration of organisms, and also the protective ozone layer. The amount of time involved is vast and may be compared with one year, in which life on earth would not appear until Christmas day.

Q9. *Make a chart (vertical or circular) to represent the history of the earth compared with a single year. Mark some of the main events shown in Figure 3.3.*

THE STRUCTURE OF THE EARTH

As the earth cooled, the initially homogeneous mass separated into layers. Again the story is complex, and speculative, but the early earth seems to have heated up, the centre becoming molten, and convection currents developed as the lighter compounds tended to rise towards the surface to form a primitive **crust** whilst denser materials such as iron sank to form a **core** which is molten (but with a solid centre produced by the pressure from above). Between the core and the crust, intermediate zones form the mantle (Figure 3.4). This differentiation also led to the release of gases from the interior (with the notable exception of free oxygen) and led to the eventual formation of the oceans and atmosphere.

- *P-waves (primary or pressure waves) shake the earth forwards and backwards. They travel quickly and are the first to arrive at the seismographs;*
- *S-waves (shear or secondary waves) pass through the earth like a wave with an up and down motion. They are slower, more powerful and cannot pass through liquids;*
- *L-waves are surface waves which travel slowly through the crust. They are last to arrive but are the most damaging.*

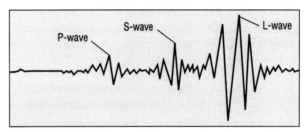

Figure 3.5 *Earthquake waves on a seismograph*

Thus the waves arriving at a seismograph can be differentiated (Figure 3.5) and using very accurate clocks their arrival times at different stations may be compared.

As the earthquake shock waves meet layers of differing density they change their directions of travel and speed up if the new layer is denser or slow down if it is less dense.

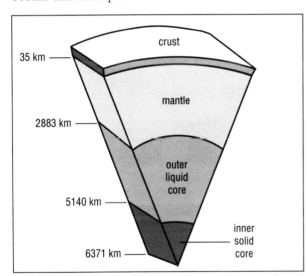

Figure 3.4 *The earth's interior*

The study of earthquake waves

How do we know what the interior of the earth is like? When an earthquake occurs the shock waves travel through the earth and are now recorded on thousands of seismographs all over the world. These shock waves are of three main kinds, each with their own characteristics:

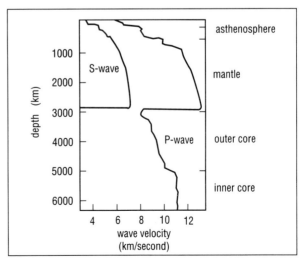

Figure 3.6 *Velocities of waves within the earth*

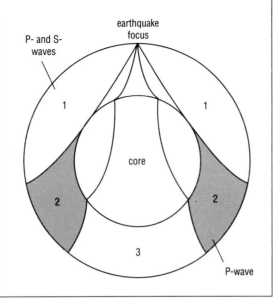

In area 1 both P- and S-waves are received by seismometers. In area 2 no P-waves are received: they are focused by the core. S-waves are not received by stations in either area 2 or 3.

Figure 3.7 *Seismic shadow zones*

> **Q10.** *Study Figures 3.4 to 3.8 and write an account of some of the ways that earthquake waves can be used to determine the nature of the earth's interior.*

Deductions based on the mass of data collected on seismographs, combined with the study of a few pieces of mantle which have been brought to the surface in lavas, and of meteorites from the break up of similar planets, have resulted in quite a detailed knowledge of the internal structure of the earth (Figure 3.8).

CRUST	Lithosphere	0–100 km	Cool, brittle
UPPER MANTLE	Asthenosphere	100–350 km	Hot, plastic
LOWER MANTLE	Mesosphere	350–2883 km	Hot but strong
CORE	Outer core	2883–5140 km	Liquid
	Inner core	5140–6371 km	Solid

Figure 3.8 *The structure of the earth (compare with Figure 3.4)*

The crust of the earth

The crust consists of all the rocks which overlie the mantle. The crust and top part of the mantle form hard slabs known as **lithosphere** and the continents are embedded in these slabs. Theoretical considerations as well as the information from seismographs and the study of actual rock samples, show that the earth's crust can be divided into:

- **oceanic crust** (sima) *i.e. the floor of the deep oceans which is thin (about 7 km thick) and made of relatively dense rocks like basalt;*
- **continental crust** (sial) *which is much thicker (averaging 33 km but increasing up to 90 km under mountains) and is composed of relatively light material such as granite.*

As can be seen from Figure 1.4 the crust is floating on the denser material of the mantle. This is the principle of **isostasy** which is of major importance in geomorphology (the shape of the land). Note that most of the crust is below sea level (just as most of an iceberg is below the waterline). It is also known (partly from studies of gravity) that the mountains have a balancing 'root' beneath them (Figures 1.4 and 3.9).

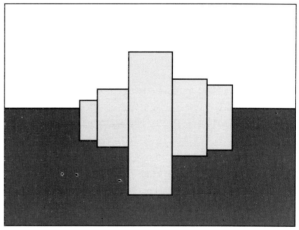

Figure 3.9 *Airey's hypothesis helps to explain the known variations in thickness of the continental crust. Wooden blocks floating in water are thought to mirror continental lithosphere floating on the denser mantle*

> **Q11.** *Study Figure 3.9 and suggest some implications arising from:*
> **a)** *the formation of ice caps over mountain areas;*
> **b)** *the erosion of mountains.*

Beneath the slab-like lithosphere (Figure 3.8) is the asthenosphere, a layer which the low velocity of the earthquake waves reveals to be weaker and partially molten. It is from here that much of the magma rises to form volcanoes, and across which the lithospheric plates slowly slide, rather like ice-floes floating across the polar seas.

SECTION 4

The ocean

KEY IDEAS

- **How are ocean floors being formed?**
- **What are the resulting submarine and surface features?**

OCEAN CRUST FORMATION

Figure 1.4 indicated that the continental and ocean crusts of the earth differ in composition and thickness because of their different origins. Ocean crust is constantly forming along ocean ridges such as the one down the middle of the present Atlantic.

Information about the ocean crust comes from seismic reflection data (Figure 3.6), from shallow cores drilled in the ocean bed, from fragments of rock brought up in volcanic lavas, from rare outcrops of ocean crust thrust up on land (e.g. in the Troodos Mountains of Cyprus) and from geophysical evidence (see page 10). The latest information is from satellite data which has already measured amounts of plate separation over a few years.

As a result of all this information it is believed that when convection currents in the earth cause continents to split apart, the crustal thinning allows magmas of basalt to rise from the asthenosphere (Figure 3.8) and reach the surface. Lavas are extruded from below and as the ocean floor moves away from the ridge the gap is filled with **dykes** (vertical sheets of magma) and the new ocean floor is slowly covered with marine sediment (Figure 4.1).

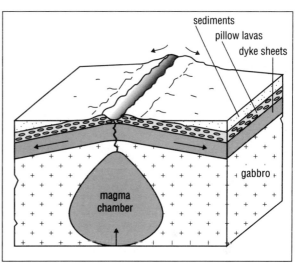

Figure 4.1 *The sea floor spreading away from a mid-ocean ridge, such as the mid-Atlantic ridge*

Q12. *Study an atlas and list the islands of the mid-Atlantic ridge. How high are the islands above the ocean floor, and how does this compare with the height of Mount Everest?*

RESULTING FEATURES

Down the middle of an ocean ridge is a **rift valley** (**graben**) (Figure 4.2), beneath which, at intervals of 10 km or more, plumes of magma rise to supply material to fill the gaps left by the separating plates. Sea water penetrating the upper crust is heated by the magma and returns in the rift valley as underwater hot springs enriched by sulphides and metals. The unusual environment for plant and animal life around these vents is currently a subject of great interest to some biologists due to the discovery of bacteria living on the sulphides and a population of unique organisms which they nourish.

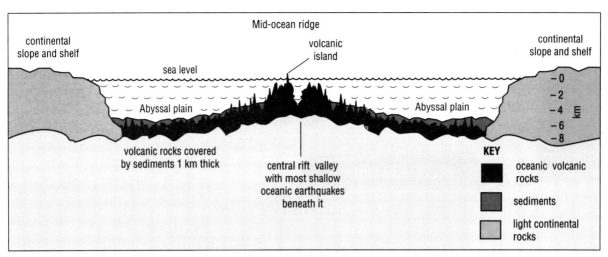

Figure 4.2 *Central rift valley of a mid-ocean ridge*

The example of Iceland

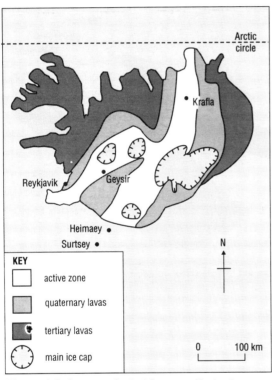

Figure 4.3 *Some geological features of Iceland*

| **Q13.** *Study Figures 2.3 and 4.3 and explain the distribution of lavas in Iceland.*

Iceland is an unusually large island on the mid-Atlantic ridge, where many of the typical volcanic features may be examined even though the absence of sea water and other geochemical clues suggest that it is not an exact model of normal sea floor spreading. The life of the islanders is dominated by their position on the mid-Atlantic ridge even more than by their climate.

One thing apparent to any visitor is the episodic nature of the movement of the plates, with separation occurring in jerks and eruptions at irregular intervals, rather than a steady slipping apart. New volcanoes may form and old ones be reactivated without much warning as in the cases of Surtsey (1963) and Heimaey (1973). There are numerous hot springs and **geysers**, including the eponymous Geysir itself, and parts of the central rift valley have pools of boiling mud and water from which rise clouds of sulphurous steam. Active volcanoes such as Krafla and Hekla are found here and at times volcanoes have erupted beneath ice caps causing devastating floods (jökulhlaups).

The volcanic characteristics affect the human geography in other ways. In this cool, wet climate the recent lavas and ashes take a long time to develop into fertile soils, so pastures are usually found on the thin veneers of sediment. The rocks are too hard to cut easily and there is no clay, so characteristic

Figure 4.4 *The eruption of Heimaey. The dormant volcano behind the town erupted in 1973 causing much damage but no loss of life*

houses are made of turf, wood or corrugated iron. Natural hot water is used to heat the houses in the capital, Reykjavik, as well as in some factories and for producing geothermal power.

Hot spots

Not all such volcanic activity occurs along the edges of the plates. Volcanoes, often forming chains, exist away from the plate margins both on continents and in the oceans. They probably form as plates move over stationary **hot spots** below the crust. The much-studied Hawaiian chain of islands illustrates this well.

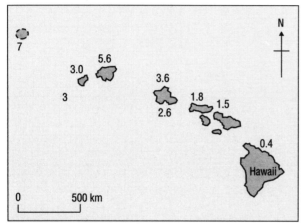

Figure 4.5 *The Hawaiian Islands with dates (Ma) of the formation of their lavas*

Q14. *Draw a section to illustrate how a plate moving across a hot spot could produce the situation shown in Figure 4.5.*

Older ocean floors

As the plates move away from the mid-ocean ridges, they cool and contract so that the ocean deepens away from the spreading centres (Figure 4.1). The ocean floor is gradually buried beneath a covering of marine sediments, some of the fine particles coming from distant rivers and from volcanic dust in the atmosphere; much being from the skeletons of the surface plankton and other organisms.

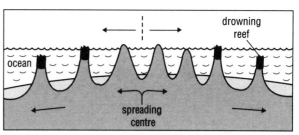

Figure 4.6 *Volcanoes with reefs (black) moving away from a spreading centre*

Volcanoes become extinct as they are carried away on the moving plates, and their surfaces may become eroded by the sea to form flat-topped sea mounts (**guyots**). In the tropics reefs of coral may form around their edges. The crust cools and contracts as it moves away from the active region and the flat tops of the guyots sink below the waves. However, the reefs may continue to grow, so forming the fringing reefs and atolls in the manner envisaged by Darwin 160 years ago (Figure 4.7). If the rate of submergence exceeds the rate of coral growth the corals drown and the reefs are submerged (Figure 4.6).

Q15. *Redraw Darwin's diagram in Figure 4.7 in three parts to show the stages of the formation of the reefs.*

As the plates separate and basalt magma rises, the huge quantities involved (estimated at about 4 km³ per year at present) displace the ocean waters so that during periods of rapid sea floor spreading, sea levels rise and the sea gradually transgresses across low lying continental areas. This appears to have happened during the Cretaceous period (Figure 3.3) and in the unusually high seas, the **chalk** was deposited over much of the British Isles (Figure 4.8). The hundreds of metres of chalk (for example 550 m in the Isle of Wight) are composed of the remains of millions of microscopic algae.

Q16. *What other causes of sea level change can you suggest?*

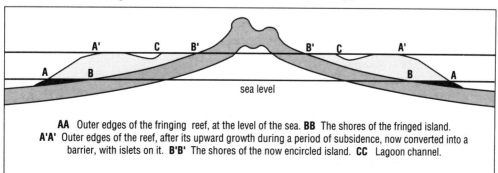

AA Outer edges of the fringing reef, at the level of the sea. **BB** The shores of the fringed island.
A'A' Outer edges of the reef, after its upward growth during a period of subsidence, now converted into a barrier, with islets on it. **B'B'** The shores of the now encircled island. **CC** Lagoon channel.

Figure 4.7 *Darwin's diagram to show the formation of coral fringe reefs, barrier reefs and atolls*

Figure 4.8 *The chalk cliffs of Dover*

One of the discoveries of the recent investigations is that the oldest part of the ocean floor, in the Pacific, is no more than about 200 million years old. This suggests that the crust being formed at the mid-ocean ridges is being destroyed elsewhere.

It is likely that, as the ocean crust cools away from the spreading ridges it becomes denser, so that after about 200 million years its weight drags it down into the mantle forming a subduction zone. Once this has been established the speed of subduction will increase so that in the end the continents will move towards each other and the ocean will close forming new, larger continents such as **Pangaea** (Figure 4.9) which had formed by the Permian period (Figure 3.3) about 225 million years ago.

By the Jurassic period (Figure 3.3) 200 million years ago, Pangaea was starting to break up again, first dividing into a southern continent (Gondwanaland) and a northern one (Laurasia) separated by the widening sea known as Tethys.

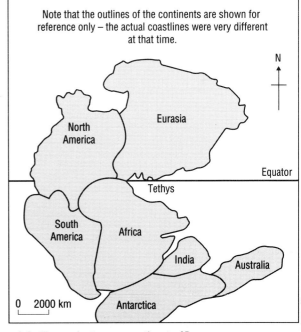

Note that the outlines of the continents are shown for reference only – the actual coastlines were very different at that time.

N

Eurasia

North America

Equator

Tethys

South America

Africa

India

Australia

0   2000 km

Antarctica

Figure 4.9 *The ancient super-continent of Pangaea*

SECTION 5

The continents

RSEINON COLLEGE LIBRARY

KEY IDEAS

- **How do the continents differ from the ocean floor?**
- **What are the causes and results of vertical movements?**
- **How do continents break apart to form new oceans?**

THE SIAL CRUST

As we have seen in Figure 1.4 (page 7), the continental crust forms rafts of sial riding on the denser plate material. The continents are thicker, older and richer in silica than the ocean floors.

> **Q17.** *Suggest why this should be so.*

The oldest rocks of the continents form **shields** and **cratons**. These probably formed before about 2000 million years ago, though some areas, such as parts of Greenland and southern Africa show dates of 3800 million years, with an exceptional age of 4200 million years being found in Australia. Some continental

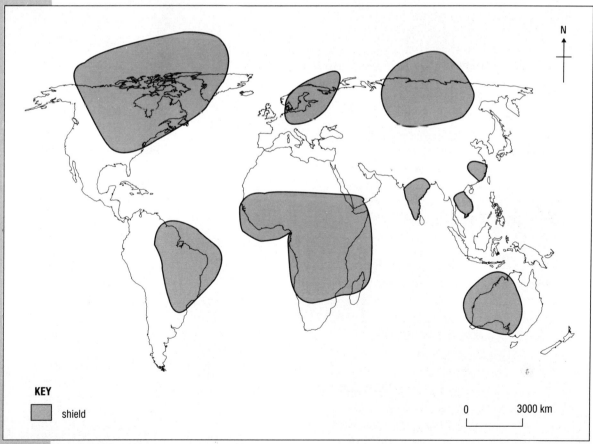

KEY

☐ shield

0   3000 km

Figure 5.1 *The major shields*

material may date from the formation of the first primitive crust but other material is formed at plate boundaries, round the past or present edges of the shields.

Isostasy

One of the most important features of the continental crust from the geomorphologists' point of view concerns the idea of isostasy. The balance of the sial crust will be upset if it is in any way loaded or unloaded, as this will result in downward or upward isostatic movements. Figure 1.4 shows the way that the sial continents float on the denser material below.

Because they float in this way the sial is thickest under mountain ranges. Figure 3.9 (page 14), shows a model which illustrates the principle involved.

The most familiar example of this was illustrated by the growth of the Pleistocene ice sheets (the Pleistocene is a division of the Quaternary, see Figure 3.3, page 12). Much of the ocean water was locked up in the ice sheets and so there was a world-wide **eustatic** fall in sea level. However, in areas near to, or actually covered by the ice, the continental crust sagged under its weight into the sima below, producing a more local isostatic rise in sea level.

> **Q18.** Draw sketch sections to illustrate the two effects on relative sea level described above.
>
> **Q19.** Rewrite the following list under two headings:
> **a)** features formed by a relative rise in sea level;
> **b)** features formed by a relative fall in sea level. Include an example of each.
>
> Ria, raised beach, buried channel, sunken forest, fjord, rejuvenated river.
> Include any other features known to you.

Figure 5.2 Raised beaches in Scotland

Complications arise because of the way that, when an ice sheet melts, there is a rapid rise in sea level, whereas the isostatic rebound of a sial continent is a much slower process. In addition, there were at least four major glaciations during the Pleistocene (three in Britain) so it is likely that the sea had started to fall before the continent had finished rising following the previous glaciation.

> **Q20.** In parts of western Britain it is possible to stand on a raised beach looking into a ria. Explain this apparent contradiction.
>
> **Q21.** Explain the origins of the features shown in Figure 5.3.

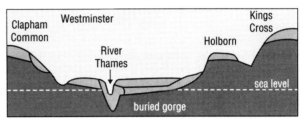

Figure 5.3 Section across the Thames Valley

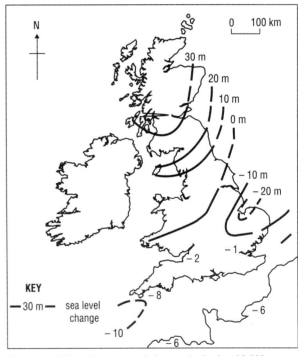

Figure 5.4 Relative sea level changes in the last 10 000 years

> **Q22.** Figure 5.4 shows the way that isostatic rebound is causing Britain to pivot, with the north rising but the south sinking. The sea is still rising at an estimated 2 mm/year. Calculate the relative changes of sea level in the next 100 years for:
> **a)** the Central Highlands of Scotland;
> **b)** Edinburgh; **c)** Hull; **d)** London.

Q23. *These seem very small changes. Calculate the time it would take for the sea level at London to rise 1 m. Relate this to the need for the Thames Barrier.*

The concept of isostasy is a fundamental part of the various theories of geomorphic cycles first developed by WM Davis in the late nineteenth century. In these theories, the continents respond to the erosion of mountains by isostatic rebound so that new potential energy is available for the rejuvenation of agents of denudation such as rivers. Although highly theoretical and subject to detailed variation and problems, the cycle theories provide one framework for considering some geomorphological processes.

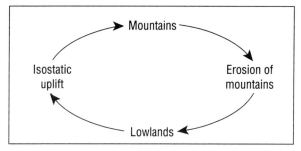

Figure 5.5 *The idea of a cycle of erosion*

CRACKING CONTINENTS

The thick continental crust is a poor conductor and so heat from the mantle may accumulate beneath it causing the plate to bulge up and crack. Faults may develop and a central rift valley form. This may widen to admit lava and ocean crust starts to form.

In this way continents break up and new oceans are created. This is happening today along the line of the Red Sea (Figure 5.6). Saudi Arabia is moving roughly north east producing a widening Red Sea and a **transform fault** along the valley of the Jordan and Dead Sea. The associated rift valley in East Africa, a system of rifts and igneous activity along a 3000 km line from the Red Sea to southern Malawi, is apparently no longer spreading and may not actually evolve into a new ocean.

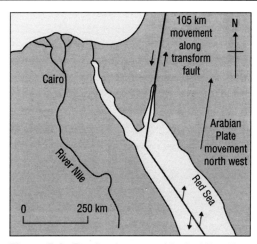

Figure 5.6 *The development of the Red Sea rift*

Figure 5.7 *The East African rift valley*

Around the edges

- **Why are most volcanoes and earthquakes found around the margins of plates?**
- **How can we explain the locations of the highest mountains and the deepest seas?**
- **Why don't all continents have these features round their edges?**

PLATE MARGINS

'The Pacific ring of fire' was a term used to describe the way that the shores of the Pacific Ocean were almost all sites of volcanoes and earthquakes (Figure 2.7, page 2), but until the plate tectonics 'revolution' the reasons for this, and for the other belts of volcanoes and earthquakes, were almost unknown. We now know that these areas of destructive plate margins are the sites of the world's most dramatic movements, where the floors of oceans descend into the earth and are consumed into the mantle.

The first type of plate margin, constructive margins is where plates are moving apart and new ocean crust is being created. This sea floor spreading is taking place along 80 000 km or so of ocean ridge, and the plates are moving away from each other at anything up to 180 mm/yr (though the Atlantic average of 20–40 mm/yr is more usual).

Q24. *If the North Atlantic is opening up at 20 mm/yr,*
a) *how long ago did it start to open? (The South Atlantic began to open about 150 ma, the North Atlantic in the late Cretaceous. Is your answer compatible with this?)*
b) *What possible sources of error may there be in your calculations?*

With plates moving outwards from the various spreading centres there must also be places where plates are meeting. It is one of the key features of plate tectonic theory that where this happens, plates are being destroyed as they slide into the earth along subduction zones.

As two plates meet, one of three things will normally happen, depending on whether one or both of the plates involved carries thick continental crust. Whichever occurs the friction involved will result in earthquakes, and the eventual melting of the subducted plate will generate magma which will rise to form volcanoes. These destructive plate margins are thus the sites of the earth's most spectacular and violent features.

Types of destructive plate margin

South America

A small earthquake in Chile has often been quoted as one of the most uninteresting newspaper headlines, partly because of the frequency of such events along the coast of South America. In that area (Figure 6.1) the Nazca Plate, made only of the denser ocean crust (sima), meets and slides under the South American plate, with its less dense sial. As well as the jerky movements which produce the earthquakes (large as well as small), the friction between the plates causes much heat, and this, together with the heat from the earth and from radioactive decay, causes the ocean plate to melt. At the same time a deep trench is formed just offshore. As the ocean plate moves under South America the sediments washed down from the Andes are scraped up against the continent to form coastal mountains.

The magma formed by the melting plate then rises through cracks in the sial to form the volcanoes of the Andes. In the end the heat may be so intense that large areas of the crust are melted to form granites below the surface (Figure 6.1).

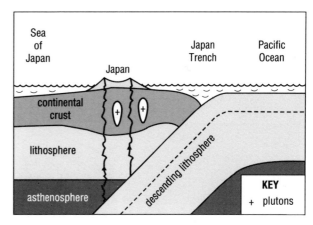
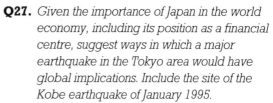

Figure 6.2 *The Japanese subduction zone*

Q27. *Given the importance of Japan in the world economy, including its position as a financial centre, suggest ways in which a major earthquake in the Tokyo area would have global implications. Include the site of the Kobe earthquake of January 1995.*

Q28. *Using the map in Figure 2.7, on page 2 and an atlas or database, draw a sketch map of the main geological features of the western Pacific from New Zealand to the Aleutian Islands. Indicate the island arcs, the deep trenches and shallow seas, and name the main ones.*

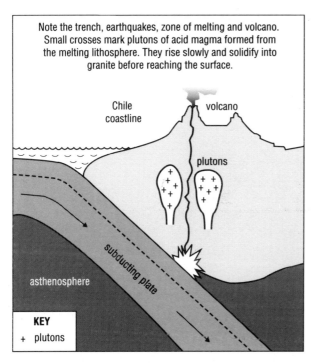

Figure 6.1 *The west coast of South America*

Q25. *Such situations exist in other parts of the world other than South America. Identify similar areas from Figure 2.7, on page 2, and list them.*

Q26. *Use Figure 6.1 to draw a sketch map of the main features of a coast where subduction is occurring.*

Japan

In geological terms Japan is a very unstable country. It is situated above a subduction zone (Figure 6.2) where ocean crust moving westwards is subducted under Japan, with similar results to those of the Andes (volcanoes, earthquakes and a deep trench just offshore). Without the subduction zone Japan would not exist, but the price to be paid is one of danger and uncertainty as it is never known when the next tectonic event will occur.

The Himalayas

Under the flags, film canisters and mint cake wrappers which one imagines cover the summit of Everest are sedimentary rocks including limestones containing marine fossils, evidence that, like other mountain ranges, the highest mountains of all were formed under the sea. The Himalayan ranges have resulted from one of the most remarkable movements of any continent, that of India. From a position in the southern hemisphere as part of Gondwanaland in the Carboniferous period, India moved rapidly northwards and eventually collided with Asia.

The result is the formation of the Himalayas, and similar though slower movements of plates have produced the other great fold mountain ranges of the world, not least the Alps, the Pyrénées and the other ranges of Europe. As two plates, each carrying continental crust, approach each other, the oceans between are filled with sediments washed down from the continents. As the two continents get closer, these sedimentary rocks are squeezed into fold mountains with the line of equilibrium through the middle (Figure 6.3). The old ocean floor is mainly subducted, though sometimes splinters break off and are caught up in the fold mountains. As before, the

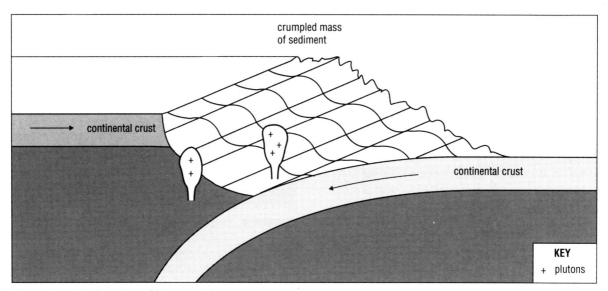

Figure 6.3 *The formation of fold mountains*

great heat from the friction, and from the heat-producing radio-isotopes in the sediments, causes melting and the formation of granites in the roots of the mountains.

Such are the events which are still producing uplift in the Himalayas though many of the detailed secrets appear to be locked in the depths of the mountains. However, it is possible to study the insides of mountains by visiting areas where similar ranges have formed in the past, but have since been eroded to reveal the details of their roots. In Scotland mountains can be seen where different types of highly metamorphosed rocks are exposed: types which require such enormous amounts of heat and pressure that they could only be formed deep in the earth. The most southerly point of England, the Lizard Peninsula, may be one of those splinters of ocean crust which was thrust up into the folds of another range which used to cross the British Isles from west to east in Permian times.

If you are going to climb Mount Everest hurry to join the queue as it is still rising! Due partly to isostatic uplift, the Himalayas are still slowly rising and this slow process has probably produced one of the more unlikely types of discordant river pattern; that of **antecedent drainage**. Several major rivers, including the Indus, Ganges and Brahmaputra, rise on the plateaux north of the main ranges, and flow straight through the mountains, often in spectacular gorges. For want of a better hypothesis it has been proposed that the rivers were established before the Himalayas were there, and that as the mountains slowly rose across their north–south courses the rivers had enough energy to keep cutting down and maintain their direction.

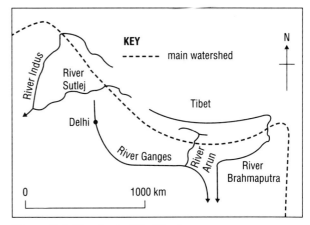

Figure 6.4 *The rivers of the Himalayas*

The three types of plate margin

There are three main types of plate boundary:

- constructive (divergent) margins exist where two plates separate and new ocean crust is formed from rising magma (Figure 4.1, page 15);
- destructive (convergent) margins exist where plates meet and these mark the sites of the major features of the earth (Figures 6.1 and 6.2);
- conservative (transform) margins are a third type and occur when two plates move past each other along transform faults (Figure 6.5). These are deep faults and as the plates grind past each other they are responsible for many earthquakes, some of high magnitude. The San Andreas Fault in California is the most notorious example of these (see Section 10) but there are plenty of others around the globe.

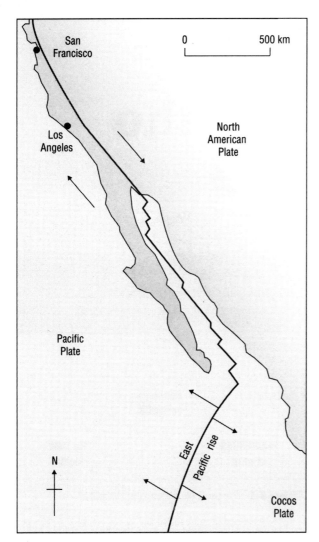

Figure 6.5 *The San Andreas Fault – a conservative plate margin*

Q29. *Where are the nearest places to Britain which are on active plate margins? Which of these are popular holiday resorts for British people?*

Q30. *Using an atlas study the continental shelves around Europe and the eastern side of North America. What sort of widths are involved? Compare these with the coasts of western America and explain the differences.*

Sea level is a very variable marker in geological terms. In the Cretaceous period the sea level was higher, allowing the chalk to be deposited on the continental shelf, whilst as recently as the Quaternary period, the sea was lower when much of the water was locked up in the form of ice during the great glaciations. During that time the shelves were mostly above sea level and much of the sediment on the European (and eastern North American) coasts is glacial or fluvio-glacial in origin, constituting a valuable resource for the construction industry. The shelves also provide important fishing grounds where bottom-dwelling species can live whilst still in the photic zone, and where nutrients from continental waters and sediments are readily available. In recent years the world's continental shelves have become even more significant as improved technology has enabled them to be searched for oil. Because of their value it is not surprising that international disputes about sovereignty of the sea bed still arise.

Passive continental margins

Of course not all the edges of continents experience such excitements. The edge of the Atlantic, off the coast of Ireland, is not an active zone of subduction like that off Chile. The reason is that the edges of continents do not always correspond with the edges of the plates.

The coasts without active plate margins are thus very different, and the British side of the Atlantic is no exception. Sea water covers over 70 per cent of the globe and some of this water lies on the shallow continental shelves which stretch for varying distances from the coastlines to about 200 m below present sea level. From there the floor drops relatively steeply down the continental slope to the deep ocean floor (Figure 6.6). The continental shelves are really parts of the continents which happen to be submerged at present.

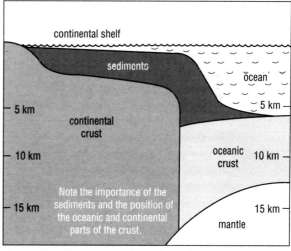

Figure 6.6 *A passive continental margin*

Much of the sediment on the shelves eventually flows over the edge and down the relatively steep continental slope, often as discrete currents called turbidity currents. These dense mixtures of sediment and water settle out at the foot of the slope to form thick deposits of greywackes (or turbidites) at the edge of the ocean floor.

SECTION 7

Rocks and the landscape

·KEY IDEAS

- **Why are some rocks harder than others?**
- **How do weathering and erosion penetrate solid rock?**
- **What are the effects of earth movements on the landscape?**

Knowledge of the movements of the plates can explain how large-scale masses of rock come to be in their present positions, but to understand the detail of the landscape we must know more about the rocks themselves.

When starting to investigate the geomorphology of an area one needs to think in terms of three groups of factors originally identified by WM Davis (1889) as 'structure, process and stage' (Figure 7.1).

The processes involved (the work of rivers, glaciers and so on) are covered in *World Wide*, the core text

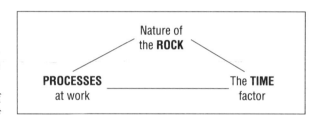

Figure 7.1 *The causes of landscape variation*

in this series, but the stage and structure factors have more of a geological basis.

HOW LONG HAVE THE PROCESSES BEEN AT WORK?

The immense amounts of time available for the processes of geomorphology have been implicit throughout this text (see page 12 especially), and so long as rocks have been exposed to denudation, they will have been under attack for that length of time. It follows that areas like the Grampians, which have been exposed to denudation for over 400 million years, will be lower than similar rocks which are younger. The fact that they are still mountainous reflects their original height, which was of Himalayan proportions. However it is not quite as simple as that, for climates have often changed and as the climates have changed, the processes at work will have changed too. It is only 10 000 years since the ice melted in Britain and ended a two million year period which was dominated by glaciation, although within

that period there were several spells when the climate was actually warmer than it is now.

Q31. *Think back to previous sections. List and explain the reasons for significant climate changes.*

One of the factors you may have identified was sea level change. The present sea level is at a very temporary position indeed, and over relatively short periods of time it has been much lower and much higher than it is now. This has not only affected the climates of different areas but has had more direct effects on the areas of marine erosion and deposition, as well as on the development and subsequent behaviour of rivers.

Q32. *If you have already studied river and coastal geomorphology, make suggestions about the sort of processes and landforms resulting from changes of sea level.*

Other changes may occur if different rock types are exposed due to overlying strata eroding away, or because rivers, for example, have cut down to expose different layers. Thus the processes may change as the rock actually being worked upon varies with time. In a somewhat similar fashion, the uplift of new mountains and plateaux will change the type of rock being attacked at the surface and cause rejuvenation of the river systems.

Q33. *Relate the width of the valley in Figure 7.2 to the different rock types exposed as the river has cut down.*

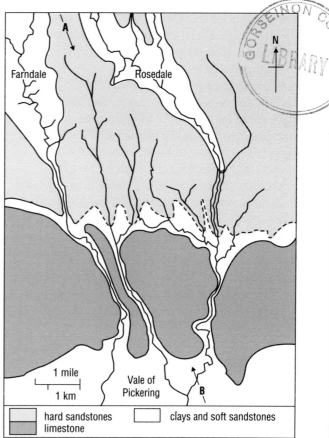

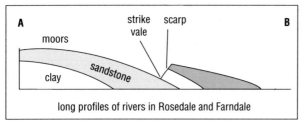

Figure 7.2 *Valley in Farndale (north east Yorkshire), cross-section and plan view*

THE NATURE OF THE ROCK

What is a hard rock?

Figure 4.8 (page 8) shows high and steep cliffs made of chalk. Chalk is one of the softest rocks there is (you can easily crush it with your foot), so why does it form hills and cliffs? On the other hand, really hard rocks, such as the metamorphic rocks of the Shetlands, may form quite low areas whilst in the case of sandstones some, like those of the Brecon Beacons, produce considerable hills whilst others form such soft rocks that they can be quarried without the need for any blasting. Rock resistance depends on three things as shown in Figure 7.3.

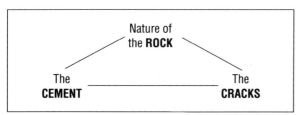

Figure 7.3 *The strength of rocks*

Rocks are made of minerals and these may be hard or soft depending on their chemical composition and crystal structure. Geologists describe the hardness of a mineral in terms of **Moh's scale of hardness** in which 1 = talc and 10 = diamond.

Sandstone is made of quartz which is 7 on Moh's scale. Quartz is a very common mineral, having crystallised from molten rock in granite and other igneous and metamorphic rocks. As well as having a strong crystal system it also has a very simple formula (SiO_2) and is less likely than other minerals to be attacked by chemical weathering processes. Thus it tends to remain, forming particles of quartz sand, when the other minerals in a rock have been altered by chemical weathering to clay and then washed away. When sand itself is deposited it will not become a rock until the individual particles have been bound together with cement, and it is this cement, more than the actual particles, which accounts for the differences between the sandstones. You can rub a piece of sandstone with your hand and usually particles will drop off. Do not be fooled by this: the quartz is a lot stronger than your hand and all that you are doing is loosening the particles from their cement which is usually much softer. Thus the nature of the cement is what really affects the hardness of the sandstone. The cement also determines the porosity of the sandstone, and because quartz is usually colourless, it is the type of cement which determines the colour of the rock. The cement may be formed at the same time as the deposition of the particles, but often it is precipitated from solutions which seep through the deposit at a later stage. The main types of cement are as follows.

● Irons of different colours (browns, yellows, reds) and strengths. Typical of these are the weak iron cements of the New Red Sandstones of Permian and Trias ages. These were formed in desert conditions and are common in many parts of Britain including the Midlands and Cheshire. Another notable iron cement of the Jurassic period comprised 20–25 per cent of the total sandstone rock in some areas of England, and this is sufficient iron to be economically usable in the Scunthorpe area, and until recently in Northamptonshire and Oxfordshire.
● Calcium cements are important in some of the sandstones in the Midlands and north east England.
● Quartz can sometimes form the cement as well as the particles, and this results in a very hard rock called orthoquartzite.

The term 'hard' rock can therefore be misleading and it is better to refer to a rock's resistance.

Q34. *Suggest the nature of the sandstone needed for: water supply, for soft water in particular; making millstones and grindstones; building stone; builders' sand; the formation of hills; and the development of moorland bogs such as those found extensively on the Pennines.*

Q35. a) *Why do soft sandstones as well as chalk form hills?*
b) *Why are the main features of glacial erosion found in areas of resistant rock rather than in soft rock areas where erosion would seem more likely?*

The sedimentary rocks

Sandstone is one of a number of clastic rocks, ones comprising fragments (clasts) held together with cement. These are classified by the size of their clasts and some of the most common are shown in Figure 7.4.

DIAMETER	TYPE	EXAMPLES
>2 mm	Rudaceous	Conglomerate Breccia
2–0.06 mm	Arenaceous	Sandstones Grit Greywacke/turbidite Orthoquartzite
<0.06 mm	Argillaceous	Clay Siltstone Shale

Figure 7.4 *The clastic sedimentary rocks*

Other sedimentary rocks are formed by organic or chemical means, and chief amongst these are the limestones which are made from calcium carbonate ($CaCO_3$), (Figure 7.5). The calcium is weathered from igneous rocks and is later extracted from the sea by organisms such as molluscs or precipitated directly, especially in areas of warm water such as those around the Bahamas.

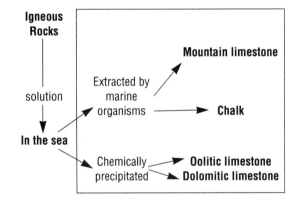

Figure 7.5 *Types of limestones*

Q36. *Economically, limestones form one of the most useful rocks. List as many uses as you can – don't forget tourism!*

The mountain limestones form the hard rocks of the Derbyshire Dales, the Mendips, the Yorkshire Dales and many other areas. Most are Carboniferous in age and are made from shell fragments or faecal pellets of marine organisms like molluscs. Being hard, these rocks tend to form hills and they are pervious rather than porous, with **joints** (cracks) produced by shrinkage as they dry out. Although divided into blocks by these joints and by the bedding planes between each layer, the actual blocks are hard, because after deposition the calcium carbonate particles recrystallised bonding the molecules together and filling the spaces. This makes them durable enough to be used in the foundations of roads and for building stone.

Figure 7.6 *Karst scenery, Craven District, Yorkshire*

Because the limestone blocks are non-porous, water falling on them runs to the edge rather than sinking in: hence the limestone pavements, swallow holes and other features of karst scenery. This limestone consists of a series of hard blocks and so it can support steep valley sides and elaborate cave systems, whereas the less resistant limestones would collapse into them.

Q37. *Find out the differences between the terms pervious, porous and permeable.*

Chalk is a remarkable rock, made of countless millions of microscopic particles from tiny marine algae. These particles of $CaCO_3$ did not recrystallise after deposition, so the rock remains porous as well as pervious, and is soft and crumbly.

The picturesque villages of the Cotswolds form part of a ridge of Jurassic rocks stretching from Dorset to the Yorkshire coasts. These **oolitic limestones** were formed chemically as particles were rolled around in warm shallow waters, collecting layers of $CaCO_3$ rather like a snowball getting larger as it rolls downhill. Like the chalk they are porous and hence they share features such as steep escarpments, dry valleys and thin soils. However, they are rather better cemented, and many of our cathedrals and other famous buildings are made of this stone.

Through much of Yorkshire and Durham, the Great North Road (now sadly known as the A1) runs along a ridge with good views of the lowlands on either side. The ridge is made of dolomitic (magnesian) limestone, which forms a thin outcrop running north and east from Nottingham to the Durham coast. It rises above the surrounding area because it is very jointed and so water passes through rather than flowing over the surface and causing erosion. The joints are a consequence of shrinkage when chemical alterations changed the rock after ordinary limestone had been deposited. This close pattern of jointing has also made it an important reservoir rock for North Sea oil.

Like some of the limestones (see Figure 7.5) coal is an organically-formed sedimentary rock, deposited in an environment which was equatorial and well-supplied with nutrients. These early versions of the tropical rain forests grew on the swamps of a large delta, evidence of the scale of geological change between the Carboniferous period and today (see Section 8 for further details). Coal is a very soft rock and huge quantities have therefore been eroded since its deposition and uplift.

The igneous rocks

Ultimately all the minerals in the sedimentary rocks can be traced back to their origin in the igneous rocks. Geologists classify igneous rocks according to their chemical composition (acid or basic) and the size of their crystals. **Lavas** such as basalt, are magmas which reach the earth's surface and cool so quickly that their crystals are less than 1 mm across. **Plutonic rocks**, such as granite, cool very slowly underground so big crystals have time to grow and these can easily be seen with the naked eye. Intermediate (hypabyssal) crystals develop where magma is intruded into cracks. Typical situations of cooling are shown in Figure 7.7.

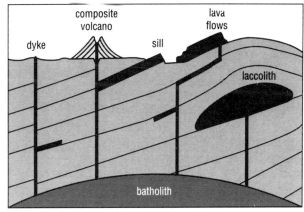

Figure 7.7 *Bodies of igneous rock*

Where sheets of magma form along bedding planes they form **sills**; where they are along cracks (usually faults) they are dykes. Because the basic magmas are less viscous than others, dykes and sills are usually made of dolerite (see Figure 7.9).

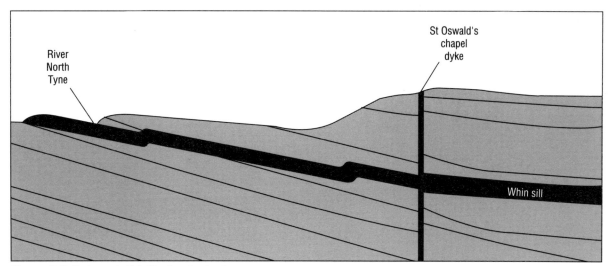

Figure 7.8 *Dyke and sill in Northumberland*

	ACID	INTERMEDIATE	BASIC
Large	*Granite*	Diorite	*Gabbro*
Intermediate	Microgranite	Micro-diorite	*Dolerite*
Small	Rhyolite	*Andesite*	*Basalt*

The more important rocks are in bold italics

Figure 7.9 *The classification of igneous rocks*

As well as extruding lavas, many volcanoes eject large quantities of **tephra**, that is fragments of solid rock smashed by the force of the explosion. Chemically they are the same as the rocks from which they were smashed and some of the main varieties are:

- **tuff**; fine particles (mistakenly called ash) which may later become slate (see below);
- agglomerate; coarse angular fragments cemented together;
- ignimbrite; a streaky rock resulting from a *nuée ardente* (an explosive mixture of gas and ash) which welds the particles together.

The shapes and sizes of the volcanoes depend on the amount and chemical nature of the magma. Basic magmas are runny and form flat sheets or gentle slopes; acid magmas are viscous and tend to produce steep-sided cones (because of this viscosity they are more likely to explode when pressure builds up inside the volcano). Composite volcanoes are formed from layers of lava and tuff: after an explosion the tephra settles quickly and is then covered with a layer of lava released by the eruption. After a dormant period the sequence is repeated. In due course the volcano will be eroded away but for a long time the resistant plug may remain (Figure 7.10).

Figure 7.10 *Le Puy volcanic plug in France*

Q38. *Using the information above, sketch sections through acid, basic and composite volcanoes.*

Layers of lava, sometimes separated by tephra, can make a series of giant steps in the landscape. This is known as trap scenery and can be seen in most areas of ancient lava flows, but particularly well in the northern part of the Isle of Skye (see also Figure 8.6, page 39) where a whole series of basalt flows of Tertiary age flowed out over each other.

Sometimes a volcano will collapse into the magma pool below forming a **caldera** (Figure 7.11). Ancient eroded examples have been identified in the Ben Nevis area, North Wales and the Lake District.

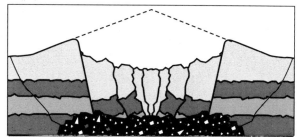

Figure 7.11 *Caldera collapse*

Large banks and other important buildings tend to be made of igneous rocks to give an impression of strength and reliability! However, not all igneous rocks are equally strong because, apart from lines of weakness in the rocks, their different minerals are not all equally resistant to weathering. On the Isle of Skye (Figure 8.6, page 39) the Black Cuillins are made of basic gabbro (see Figure 7.9) whilst the neighbouring Red Cuillins are made of acidic pink granite. Both mountain areas have been uplifted and both have experienced glacial and other attacks, but the Black Cuillins are the ones which attract climbers because of their crags, gullies and screes whereas the Red Cuillins are rounded and safer. The reason is that not only have the Black Cuillins been intruded by sheets of intermediate magmas, but the minerals of the basic gabbros are less resistant to weathering and erosion than those of the granites.

a)

b)

Figure 7.12 *The Cuillins of Skye a) Red Cuillins b) Black Cuillins*

Even granite, which is generally more resistant than most, can be broken down by chemical weathering in circumstances where the weaker minerals are subjected to intense chemical attack. This happens particularly in the tropics, where chemical action is more intense, but also where waters are especially acid, or where solutions from within the earth rise and attack the rock (the china clay deposits of Cornwall were formed from granite in this way).

The metamorphic rocks

These are rocks which have been altered by heat and/or pressure. Only those most likely to be encountered by the geographer are mentioned here.

Thermal (contact) metamorphism occurs when the country rocks are intruded by magma such as granite which 'cooks up' the rocks in the surrounding **aureole**. Common examples occur when:

● limestone changes to marble;
● sandstone changes to metaquartzite;
● shale changes to hornfels.

These are all hard rocks but because of their mode of formation they do not cover large areas.

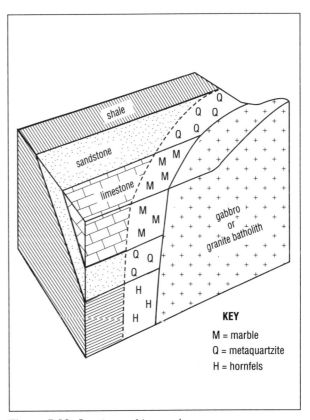

Figure 7.13 *A metamorphic aureole*

Regional metamorphism occurs when rocks are changed inside the earth's crust by extremes of pressure and usually heat as well. The three main types are:

● slate (the lowest grade);
● schist;
● gneiss (the highest grade).

Slate is formed when fine particles of argillaceous rock are compressed so that the minerals become orientated in one direction forming slaty cleavage along which the slates can be split into fine sheets.

Most British slates like those of North Wales, originated as marine muds whereas those of the Lake District were originally volcanic tuff. The Welsh slate had better cleavage and was used all over the country after the coming of the railways. The Lake District slate cleaves less well and although Cumberland Green Slate is still used for roofing, its small blocks are the traditional building stone of the Lake District. The National Park Planning Board insists on its use for new building in that area and so the quarries there are flourishing, in contrast to the sad abandoned slate quarries of North Wales.

Schist and gneiss were subjected to even greater heat and pressure in the roots of mountains and so are more altered than slate. Regional metamorphism on this scale occurred in Scotland and most of the Highlands and Islands consists of these two rock types. The fact that they still form mountains despite being 500 million years old reflects their hardness and also the huge scale of the mountain chain which has since been eroded.

Lines of weakness

If rocks were really solid with no cracks or pores in them, weathering and erosion could barely scratch the surface and water would be unable to penetrate to carry the solutions for chemical weathering and to cause freeze-thaw. Changes would only be made in the top few millimetres of the rock and there would be little sediment and few plants as there would be no soil.

However some rocks are porous, and all have a variety of cracks, some more than others.

● Bedding planes are formed between beds of sediment (and of lavas). Most start being roughly horizontal although they may be tilted later.
● Joints are usually at right angles to bedding planes and are formed in different ways:

● thermoclasty splits the rock as it heats and then cools;
● contraction joints form when a rock cools or dries out;
● tectonic joints form during orogenies;
● dilation joints are formed by release of pressure.

All of these are important in allowing the agents of denudation to attack inside the rock as well as on the surface. Dilation joints (pressure-release joints) deserve special mention as they are important in erosion as well as weathering. When the country rock is eroded from the surface of a granite **batholith**, the removal of its weight causes the top layer of the granite to rise so that joints form parallel to the surface.

Denudation then removes this upper layer causing further release of pressure and so the process starts again. The erosion is thus self-perpetuating and an example of **positive feedback** in action.

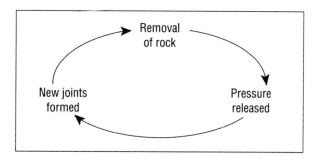

Figure 7.14 *The formation of dilation joints*

Similar effects can be seen in a variety of situations including around the edges of quarries and at the top of sea cliffs, where the cracks formed by the dilation joints are often clearly visible.

Q39. *Draw diagrams to illustrate the effects of dilation joints on the erosion of:*
    a) *sea cliffs;*
    b) *a glacial valley or corrie.*

LANDSCAPE FEATURES

Faults and the landscape

During orogenies when earth movements are too fast, or the rock is too cool or hard, the rocks will suffer brittle deformation and will fracture, thus producing faults. The three main classes are shown below and in Figure 7.15.

● Reversed faults occur when one layer is thrust over another by pressure from one or both sides

and is thus most likely to occur when plates come together during an orogeny. (A reversed fault with a particularly low angle is a thrust fault.)
● Normal faults are caused by the weight of rocks pushing down and outwards or by crustal extension away from the fault, and so are more likely to occur when plates are not pushing together, or when they are moving apart.
● Wrench (tear) faults occur when the two sides of

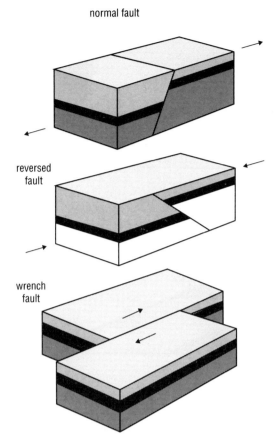

normal fault

reversed fault

wrench fault

Figure 7.15 *Types of fault*

the fault slide past each other. They are in effect smaller versions of transform faults (see Section 5).

The features of rift valleys (graben) are formed when the ground between two faults drops, whereas horsts occur when an area is left standing between two faults.

Most British faults originated in the Caledonian or Hercynian orogenies (Figure 3.3, page 12) though some have moved a little since then.

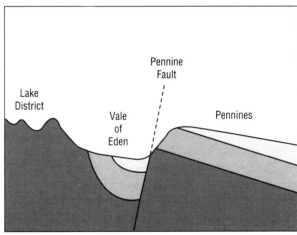

Figure 7.16 *The Pennine Fault*

The most common surface manifestation of a fault is a fault scarp, such as the Pennine Scarp which runs for some 40 km along the western edge of the North Pennines, overlooking the Vale of Eden and the Lake District (Figure 7.16).

The main rift valley in Britain holds the lowlands of central Scotland – the Midland Valley – and in contrast to the Highlands to the north and the Southern Uplands to the south, this area has fertile fields, long estuaries and better weather, whilst beneath the surface, deposits such as coal and oil shale have a long history of exploitation. It is therefore no surprise that this is the most economically important part of the province containing the largest cities and most of the population.

In areas of soft rock such as those of eastern England, rivers do not have too much trouble cutting down wherever they are. However in areas of harder rock the presence of fault lines will often encourage streams and rivers to pick out these lines of weakness. Rivers naturally bend or meander in their courses so that if you see a natural stream or river following a straight course then it is likely to be following the line of a fault. A good example of this is the Great Glen, a wrench fault whose sides have been displaced some 160 km. Rivers eroded this line of weakness and their courses were eventually followed by glaciers, leaving the deep straight valley through the Highlands with its chain of ribbon lakes which can be seen today.

Q40. *Use an atlas map to reconstruct Scotland as if the Great Glen fault had not occurred.*

A further important aspect of faulting is the way that faults may be followed by intrusions from below. Sometimes the cracks are filled with magma to form dykes, and at other times valuable materials such as lead and copper solidify in veins of mineral ore.

Folds in the landscape

When two plates come together the most obvious result is the compression of the rocks, and this plastic deformation has created all the great fold mountain ranges of the world. There are numerous technical terms to describe the different sizes and shapes of folds, but the basic ones are anticline and syncline (the latter being a downfold like a sink).

Notice from Figure 7.17 that the youngest rocks are in the middle of the synclines and the oldest are in the middle of the eroded anticline. The left hand syncline is asymmetric.

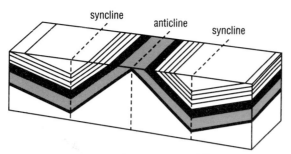

Figure 7.17 *Two synclines and an anticline*

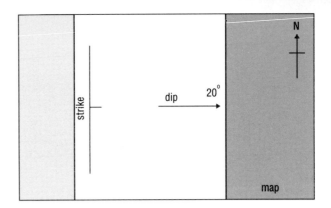

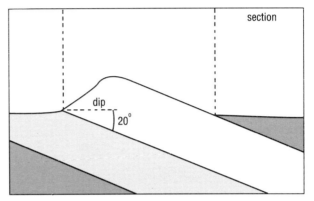

Figure 7.18 *Dip and strike*

Two useful terms for describing the behaviour of folded rocks are dip and strike (Figure 7.18). Dip is the maximum angle from the horizontal (and its direction) down the slope of the rock, for example: 20° NW (such a dip would be shown on a map as ⟍20°). Strike is the direction of a horizontal line along the bedding plane of the dipping stratum. For example the chalk of the Chilterns has a strike which is SW to NE, whereas in the Yorkshire Wolds it is SE to NW. Inspection of Figure 7.18 will show that strike is at right angles to the dip.

So far, so good: the Derbyshire Pennines are an anticline; and London is in a syncline (the London Basin). However, as Figure 7.19 shows, Snowdon is at the bottom of a syncline, whilst the large valley of the Weald is along the site of an anticline. This may seem rather unusual, but the fact is that there are numerous similar examples even if they are on a less impressive scale. Because of the way that these anticlines form valleys and the synclines hills (the opposite of what one would expect) the situation is known as inverted relief.

The reason for this is that when the crust is stretched the outer part has further to go (like the outside lane of a running track) so that joints form along the crest of an anticline and over long periods of time they are sites where weathering and erosion can most easily attack the rocks. In the case of the synclines, the lower layers are compressed thus making them more resistant to erosion.

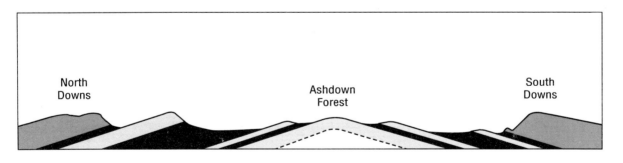

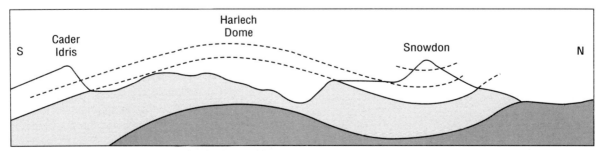

Figure 7.19 *Inverted relief*

SECTION 8

Britain's scenery

KEY IDEAS

- **When were England and Scotland divided by an ocean?**
- **Why do many of our rocks suggest that past climates were very different?**
- **Why do we have old volcanoes and other igneous rocks in Britain?**

GEOLOGICAL TIME

The vast amount of geological time is divided into periods and these are shown on Figure 3.3 (page 12). For most of earth's history, the **Pre-Cambrian**, relatively little is known of the development of any area, but for the rest of geological history, the phanerozoic, increasing detail is known. Some of the major events in the geological history of Britain are outlined below. *(It is recommended that a coloured geological map of the British Isles is to hand when studying this section.)*

An ancient shield

The oldest rocks in Britain are in north west Scotland. Pre-Cambrian metamorphic rocks form the lowlands of the Outer Hebrides and the north west mainland, with spectacular hills of ancient sediments (the Torridonian Sandstone) rising from the low ground (Figure 8.1). These ancient rocks formed part of an extensive shield which stretched westwards across what is now Canada, and it was only 40 million years

Figure 8.1 *The Torridonian sandstone mountains of north west Scotland, resting on the ancient rocks of the Pre-Cambrian shield*

ago, with the opening of the North Atlantic ocean, that North America departed westwards leaving this fragment of the Canadian shield in the North West Highlands.

The Andes of Britain

The high mountains of Britain, those of Scotland, the Lake District and Snowdonia, are mainly in the west of the country. Most are rugged as well as high and they show the effects of glaciation upon the hard igneous rocks. Others, such as those of central Wales and the Southern Uplands, are less resistant, being made from hard sediments called turbidites (or greywackes). These were usually smoothed by the ice sheets, except where corries and U-shaped valleys bit into their sides. Both these rock types, as well as other less extensive varieties, were formed because a great ocean, perhaps the same size as the present North Atlantic, separated England from Scotland during the Lower Palaeozoic Era (see Figure 8.2). This ocean, known to geologists as **Iapetus**, also split the present North American continent, as the palaeogeography of that era was very different from that of the present day. There is now no direct sign that the ocean existed, since it closed at the end of the Silurian period and the two sides met roughly along the present English–Scottish border, creating mountains at least as high as the present Himalayas.

> **Q41.** *Draw a labelled sketch section to relate Iapetus to Figure 4.2 on page 16.*
> **Q42.** *Study Figure 8.2 and list some of the key present-day areas which were:*
> **a)** *on the northern side of Iapetus;*
> **b)** *on the southern side.*

The history of the Iapetus Ocean can best be explained in three phases.

Phase 1

During the late Pre-Cambrian and Cambrian periods the ocean was opening, with sea floor spreading taking place, as it does now along the mid-Atlantic Ridge (see Section 6). **Pillow Lavas** and cherts were formed in the deep ocean and although most of these were subducted when the ocean closed, a few examples can be found in Anglesey and near Girvan in south west Scotland.

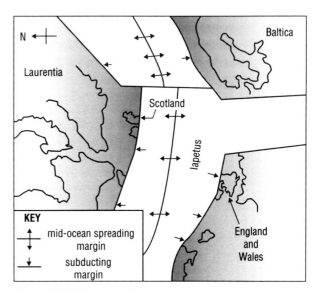

Figure 8.2 *The Iapetus Ocean*

Phase 2

For most of the Ordovician period, subduction of the ocean plates was occurring in much the same way as happens round the edges of the Pacific Ocean today. There must have been severe earthquakes, and there were certainly volcanoes, as the melted material from the subduction zone rose through the sial crust to form the volcanic rocks of Snowdonia and the Lake District. A great variety of lavas and tephras is well displayed to walkers and climbers in all of these areas, and the whole basis of the tourist industry lies in the effects of the Quaternary glaciation on these Lower Palaeozoic igneous rocks! For part of the time the central Lake District formed a great caldera, with the tuffs and agglomerates being deposited in the water of a lake, forming characteristic **lacustrine** features which can be clearly seen today. The main volcanic rocks were andesites, so-called because they form much of the Andes. Britain in the Ordovician was in a similar position with respect to a subducting plate margin, so similar rocks were formed (see Figure 6.1, page 23).

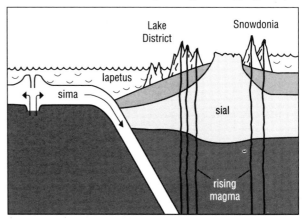

Figure 8.3 *Iapetus in the Ordovician period*

Phase 3

During the Silurian period, the ocean started to close and, helped by a lack of vegetation (land plants were only just starting to evolve), enormous quantities of sediment were being washed down from the new mountains into the edge of the ocean creating great pile of turbidite (or greywacke) which today form the rounded hills of central Wales and the southern part of the Lake District (known as Furness). As the ocean plate was subducted under the northern continent, these new sediments were scraped up to form the Southern Uplands of Scotland. The closure of Iapetus may be likened to the formation of the Himalayas today (Section 6) and the resulting mountains (known as the **Caledonian mountains**) may have been at least as high, since 500 million years of erosion have lowered them significantly. The folding pressures were intense, for not only is there a 30 km strip running across Scotland from south west to north east, which has been turned completely upside down, but in many areas the heat and pressure led to the metamorphism of the deposits. At a late stage the melting of the crust led to the intrusion of many of the Scottish granites. On the southern side of Iapetus the results of the Caledonian orogeny were less severe,

but pressures were great enough to produce the fold mountains as well as the famous slates and granites of the Lake District and Wales.

As the ocean closed, the continents moved towards each other from the north west and from the south east, so the trend of the Caledonian mountains runs south west to north east and this direction can be traced in the mountains of New England, the British Isles and Scandinavia. As the two continents came together not only fold mountains, but also large fault systems were created, especially in Scotland. There is a smaller version of the San Andreas fault in the wrench fault (tear fault) which forms the Great Glen (modern estimates suggest that the north west side of this fault slid south westwards for 160 km relative to its south east side). In the far north west of Scotland pressure from the south east forced the main rocks of the Highlands over the ancient shield, along the line of the Moine thrust fault. The Midland Valley of Scotland, with its arable lowlands, harbours, cities and coal deposits, owes its lowland origin to the formation of a rift valley between the Highland Boundary Fault to the north, and the Southern Uplands Fault along the southern edge.

BRITAIN'S CHANGING CLIMATES

The northward drift

Since the excitements of the Lower Palaeozoic, Britain has been drifting slowly northwards with a recent turn to the east. This resulted in a variety of different climates as the area crossed the climate belts of the world in order, from south to north. In the Devonian, Britain was in the arid and semi-arid area along the Tropic of Capricorn and great thicknesses of the semi-desert deposits of Old Red Sandstone were eroded from the Caledonian Mountains, to be deposited in basins within the mountains in south east Wales, and in the Midland Valley and the far north east of Scotland. At this time the south of Britain was beneath the sea where limestones and other marine rocks were being formed.

The Carboniferous period which followed was probably the most economically important in Britain's geological history. By now it was very near to the equator and a warm shallow sea covered much of the country. In this sea, corals and other marine animals converted the calcium in the sea water to the great thicknesses of limestone which have considerable economic importance.

Q43. *Make an annotated list of as many uses of Carboniferous rocks as you can think of.*

At this time, Britain was still attached to North America and Greenland and from there an enormous delta was spreading southwards, gradually filling the shallow Carboniferous sea. The material of this delta forms the **millstone grit** which (apart from millstones) provided the soft water for the textile industry, the grey houses of the Pennine towns, and the breezy but boggy uplands of the Pennine Way from Kinder Scout northwards to Scotland.

Much more significant were the swampy areas on the top of the delta. Equatorial forests flourished there, nourished by the minerals brought down by the great river, and although the species of trees were different from those of today's rain forests, they grew in similar profusion. When they fell down they lay in the delta pools and swamps which, being almost devoid of free oxygen, encouraged little action by the usual decomposers of organic matter, the bacteria and fungi. Under the weight of its own sediments the delta sank and the wood was covered by marine sands and shales, the delta advanced again and another layer of trees grew and died. This process repeated itself dozens of times, and each layer of wood was eventually changed by pressure and a little heat, into a coal seam.

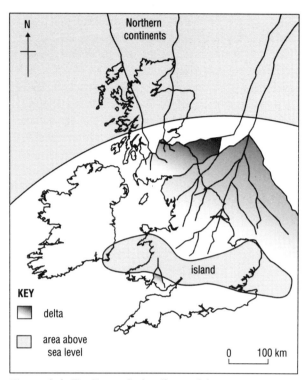

Figure 8.4 *The Upper Carboniferous delta*

At the end of the Carboniferous period, Africa moving northwards collided with Europe, producing mountains in southern Britain and much of the rest of Europe in the way described in Section 6. The west–east folds of south west Eire and the syncline of the South Wales coalfield were formed, as was the Pennine anticline. In addition, granite batholiths were intruded: the best-known being the one underlying the south west peninsula of England which is now exposed to form Dartmoor, Land's End, the Scilly Isles and the other moorlands of south west England (except for Exmoor).

> **Q44.** *Using a map from an atlas or database study the location of the main coalfields of Britain. Note that the fields on either side of the Pennines are part of the Pennine anticline. What difference might it have made to the economy if this had been a syncline, like that of South Wales?*

Continuing the northward drift, Britain then experienced Sahara-like desert conditions along the Tropic of Cancer. Flash-flood deposits from **wadis**, and large sand dunes were all formed and can be

seen in the orange-red building stones of the Midlands, east Devon, the Vale of Eden and many other areas. In the east a shallow sea evaporated to leave the various salt deposits which form the basis of the Teesside chemical industry. In Cheshire a great salt lake formed, the salts of which are spread on British roads, as well as supplying the second chemicals centre around the head of the Mersey estuary.

Figure 8.5 *Desert dunes in Cumbria*

> **Q45.** *Using a geological map from an atlas or database, draw a map of England to show the areas of the Permian and Triassic building stones (the New Red Sandstone) and mark the Teesside and Cheshire chemical industries.*

The Jurassic and Cretaceous periods

During the Jurassic and Cretaceous periods, Britain was again usually covered by shallow warm seas in which limestones were commonly deposited. These formed the significant oolitic limestones of the Cotswolds, the hills of the East Midlands, the famous Portland Stone of the Dorset coast, which is much used in the buildings of London (all these are of Jurassic age), and the great thicknesses of chalk formed in the upper Cretaceous from sub-microscopic remains of marine algae.

> **Q46.** *Map the main limestones of Britain.*

Thick deposits of clay were also deposited and these are used for brick manufacture, especially in the Peterborough, Bedford and Oxford areas. One other important event in the Jurassic was the formation of oil. In a shallow warm sea there occurred algal blooms in which vast quantities of plankton died.

Their bodies rained down onto the sea floor and were buried and preserved, to be converted by heat and pressure into the oil of the northern part of the North Sea Basin.

The Cainozoic era – North America becomes independent

The Cainozoic era began quietly enough with the deposition of sands and clays over what is now southern England. However, since at least the Jurassic period, convection currents within the earth's mantle had been starting to tear the ancient continent apart.

Below the present North Sea the crust had thinned, and in places collapsed, producing the pattern of faults which had helped to trap the North Sea oil. By 60 million years ago a north–south line of weakness had been established down the west coast of Scotland and as far south as Lundy in the Bristol Channel. Along this fracture a series of volcanoes erupted producing lava flows and tephra, and beneath the surface other igneous rocks such as granite and gabbro were being placed. A good example of this can be found in Skye (Figures 7.12, page 31 and 8.6) where the rounded hills of granite,

Figure 8.7 *The Giant's Causeway – columns of basalt formed in the Tertiary*

the Red Cuillins, contrast with the rugged Black Cuillins which are made of gabbro, a rock which chemically weathers more easily and is crossed by faults, dykes and other lines of weakness. In Northern Ireland a vast lava flow, the Antrim Plateau, occupies almost half the province, and its northern edge includes the characteristic basalt columns of the Giant's Causeway (Figure 8.7).

However, in the end the North Atlantic did not open up along that line (if it had, Ireland would now be part of North America!). The sea floor spreading began a little further west and since that time America and Britain have been moving apart at about 2 cm/year.

Later in the Cainozoic, further pressure on Europe from the south folded up the Alps, the Pyrénées and other mountains of the Alpine chain. Britain was too far north to experience much folding though the 'outer ripples of the Alpine storm' did produce the synclines of the London and Hampshire Basins, and the anticline of the Weald, now famous as an example of inverted relief.

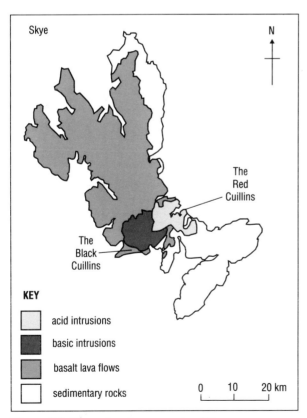

Figure 8.6 *The various igneous rocks of Skye*

> **Q47.** *Use a geological map to sketch a cross-section from the Chilterns, south east to London and then south through the Weald, to the coast near Brighton. Label the escarpments of the Chilterns and the North and South Downs, the Weald anticline and the London Basin syncline. What are the implications for the water supply of London?*

Since that time the dominant processes affecting Britain have been ones of erosion, notably those of the Quaternary glaciations between two million years ago and 8300 BC. Geologically speaking we continue to drift slowly eastwards and await the next glacial advance.

GORSEINON COLLEGE LIBRARY

SECTION 9

Applications of plate tectonics

KEY IDEAS

- **How have plate tectonics influenced world climates?**
- **What economic resources are directly determined by plate activity?**
- **Why do people live in geologically hazardous areas?**

PLATE TECTONICS AND CLIMATE

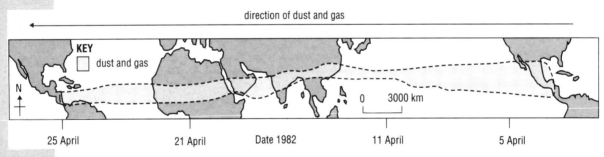

Figure 9.1 *The spread of dust and gas from El Chichōn's eruptions, 28 March and 4 April 1982*

Volcanoes and climate

In late March and early April 1982 a volcano in southern Mexico, called El Chichōn, erupted violently three times, killing 200 people and shooting clouds of gas and dust more than 25 km upwards into the stratosphere. Using satellites, the resulting clouds of material were followed as they moved westwards round the earth, and within three weeks they had completely circled it (Figure 9.1).

In June 1991, the volcanic Mount Pinatubo in the Philippines, having lain dormant for 600 years, erupted with the loss of about 400 lives. Twenty million tonnes of volcanic dust were generated, and whilst some settled on the local area, blanketing the fields and causing the collapse of house roofs, much

of it was carried into the atmosphere. As a result it is thought that global temperatures were reduced by almost 0.5°C and it has been suggested that global warming (if it exists) may have been delayed for several years.

Very large volcanic eruptions increase the dust and aerosol content of the upper atmosphere and can affect the weather for long periods. However, as with all forms of weather forecasting and investigation, there are so many different variables affecting the atmosphere that it is hard to isolate those caused by volcanoes. The local effects are clear enough; the rapid upward movement of air draws in violent winds, and moisture condenses rapidly forming heavy muddy rain and often thunderstorms.

Figure 9.2 *The eruption of Mount Pinatubo, 1991*

On a global scale the very small particles injected into the stratosphere may remain there for months or even up to seven years. Sulphur dioxide picks up water and oxygen to form fine aerosol droplets of sulphuric acid and these along with the tiniest particles of dust can cause colouring and prolonging of sunsets around the world as well as absorbing some of the sun's shortwave radiation, thus reducing the amount reaching earth. This is thought to cause some global cooling, especially as the layer of dust and aerosols does not prevent the longwave radiation of heat from leaving the atmosphere. There is now considerable evidence to suggest that periods of volcanic activity correlate with cooler weather conditions, whereas periods of less frequent eruptions (such as earlier this century) coincide with periods of global warming.

Q48. *Carry out research on the layers of the atmosphere (e.g. Section 4 of* World Wide, *the core text). Draw an annotated diagram of the atmosphere and stratosphere to illustrate these effects.*

Other implications of these events are less clear because of the large number of variables involved in climatology. For example there is much speculation about the effects of such eruptions on El Niño, a phenomenon concerning ocean currents and temperatures in the equatorial Pacific. At times the tropical waters of the eastern and central Pacific become unusually warm, with important effects on ocean currents and the weather. Changes in El Niño have been blamed for serious climate variations in western South America as well as floods in California, droughts in Australia, and possibly even more widespread events such as desertification in Africa and the global extremes of meteorological events in years such as 1987. It may be that El Niño is also affected by the heating effects of new lavas, as ocean crust is formed along the East Pacific Rise spreading zone; but evidence does suggest that it is likely that its erratic behaviour may be linked to some of the world's major volcanic eruptions, though as yet the verdict is 'not proven'.

Q49. *Find out more about El Niño and its possible effects on world climates.*

Climates and continental drift

As continents have broken up, reformed and moved to different parts of the globe the distribution of land and sea will have been affected and this in turn will have led to changes in climate.

● Because Britain has drifted northwards from a position near the South Pole in the late Pre-Cambrian period, we have experienced a great variety of climates during the Phanerozoic, starting with glaciations, and then moving northwards through the different climate zones (including equatorial and desert) to our present position. The evidence for past climates is preserved in the rocks of our geological column.

Q50. *What sort of evidence is provided for past climates by the geological column? List some suggestions with as much detail as you can.*

● When large continents existed the climates of their interiors would have been more extreme: they would have had greater ranges of temperature and they would have been more arid. Such situations have existed at various times in the past, notably in the case of Pangaea (Figure 4.9, page 18). This super-continent had been created by the Permian period, and its subsequent break up led to the eventual formation of the present continents. It is possible that the extremes of climates experienced in the interior of Pangaea contributed to the extinction of some terrestrial animals and plants.

● These hostile characteristics may have been increased by the formation of mountain ranges as the oceans closed, so that the continental interiors were even more isolated from the maritime effects of the oceans. Mountain ranges can also affect air movements, both in the upper atmosphere and nearer the ground. These in turn would have produced different movements of air masses, as well as affecting wind speed and direction, the behaviour of jet streams and the movement of depressions. Thus the climates of the past have been different, even in areas now occupying similar latitudes to the present continents.

● Relatively small changes in the levels of the oceans can drown or expose large areas of continental shelf affecting the climates of nearby areas.

Q51. *Discuss in more detail, the climatic results of the formation of a very large continent.*

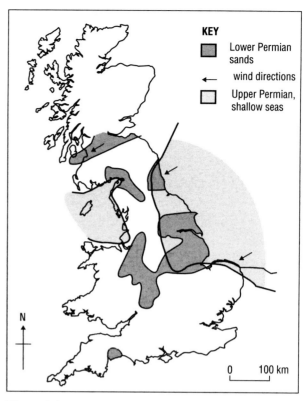

Figure 9.3 *Britain in the Permian period*

● The position of a land mass at or near one of the poles would encourage the growth of large ice caps and 270 million years ago part of the super-continent of Pangaea (see above) was at the South Pole. The evidence for this can be found in areas which are now tropical (Figure 1.3, page 6).

● The changing configuration of the ocean basins can allow other influences to alter the climate. Ocean currents are a major means of transferring heat from the tropics to the poles and warm currents entering the Arctic and Antarctic are important in determining the temperature and, more importantly, the precipitation of those areas. Although the immediate causes of the current glacial advances and retreats are probably connected with **Milankovitch cycles**, the position of the continents, the opening and closing of oceans, and the creation and destruction of submarine ridges are all necessary to ensure that conditions for the formation of ice sheets are in place.

Q52. *Draw maps to compare the distribution of glaciers and ice sheets within the Arctic and Antarctic Circles, and explain the differences between them.*

Q53. *For each of the examples of climate change above, expand on the likely causes and effects of the suggested changes.*

ECONOMIC RESOURCES AND PLATE ACTIVITY

Mineral deposits and tectonics

The discovery of economically viable minerals has always been a hit and miss affair, relying more on luck than theoretical knowledge of where they might be found. With the increasing understanding of how the earth works it should be easier to more accurately locate areas worth exploiting for particular minerals.

Magmatic separation is a process which occurs in plutonic igneous rocks as they cool. In southern Africa, for example, a large mass of gabbro contains deposits of magnetite and chromite, which separated early in its cooling and sank to the bottom of the magma chamber. Now that this ancient rock is exposed the minerals are available for exploitation.

Hydrothermal deposits are formed when hot solutions of brine dissolve minerals from rocks or from cooling magmas, and then precipitate them elsewhere, in cracks (faults or joints), or on coming into contact with cooler waters. Fluids of this type were probably responsible for the mineral veins found along the faults and joints of rocks in parts of Britain where heat was provided by the igneous intrusions such as the Cornish granite and that which underlies the Lake District and the North Pennines.

The Red Sea is situated on a spreading axis and the heat from below is causing water to circulate through the igneous rocks dissolving metals and depositing them again in the colder sea waters.

Along constructive plate margins, underwater hot springs at depths of 3 km or more eject water at 350°C, and this contains solutions of minerals such as iron, zinc and copper which are precipitated onto the sea floor. These deposits move away from the ridges and are subducted at destructive margins, perhaps to rise again in the marginal mountain chains and island arcs. However in a few cases, splinters of ocean floor are scraped off and thrust up onto land where the minerals may be exposed. The Troodos Mountains in the island of Cyprus (named for its copper deposits) is one of these areas, and its copper sulphides, as well as other minerals, have been mined since the Bronze Age.

The faulting and heat flows at plate margins, together with the varieties of igneous rocks to be found there, make these areas, both ancient and modern, important for mineral exploration. So far, however, research has tended to link particular minerals with igneous rock types, rather than with particular plate tectonic situations.

> **Q54.** *Use a world map of mineral deposits (from an atlas or database) to compare their distribution with:*
> **a)** *plate margins;*
> **b)** *cratons (Figure 5.1, page 19).*

> **Q55.** *The floors of the oceans are covered with valuable mineral deposits, often in the form of nodules, from a few centimetres, up to the size of footballs. What would be the economic and political problems involved in their exploitation?*

Geothermal power

The vast amount of heat within the earth would seem to represent a major asset, but so far our capacity to use it is rather limited. The best hope is to use hot water and steam to provide natural heating, or to generate electricity. However in most parts of the world the necessary hot rocks are too far beneath the surface for economic exploitation and it is only in volcanic areas that water at the required temperature is readily available. Other problems include the need for the water to be trapped in pervious rocks, and technical problems such as the high pressure of the steam and the tendency for hot solutions to attack the pipework and so to require careful disposal. In a few cases, notably at Larderello in Italy, steam does not come from active igneous areas but from an anticline above a still hot granite mass. Experiments are under way in Cornwall and elsewhere to produce hot water by passing cool water down boreholes in dry granite into depths where the rocks are hot enough. The water then passes through the hot rocks and is drawn up a parallel borehole. With the various concerns about the use of the world's fossil fuels, and the search for 'clean' alternatives, we can expect greater research efforts in this area in the future. At present geothermal power is being used or developed in the following countries (in order of planned production): the western USA; the Philippines; Italy; New Zealand; Japan; Mexico; Turkey; Guadeloupe; eastern CIS; Iceland; El Salvador; Chile; and Taiwan.

> **Q56.** *Compare the countries in this list with the maps of plate margins. Describe your conclusions and comment on the implications.*

A hazard is a situation which *might* cause people harm, whereas a risk is a chance that people *will* be harmed by the hazard. Volcanoes and earthquake zones therefore constitute hazards but you could avoid the risk of death or injury by staying away from them.

> **Q57.** *Which of the common destinations for British holiday-makers present geological hazards? Why do so many people take these risks?*
>
> **Q58.** *Use a map from an atlas or database to identify the major cities of the world and make two lists according to whether or not they are in areas which are geologically hazardous.*

Volcanic hazards

It has been estimated that 85 000 people have been killed by volcanoes this century. Apart from dangers from the blast of the eruption, there are many hazards associated with volcanoes. **Volcanic bombs** (blobs of semi-molten lava) which shower down, have killed many people, and fine tephra will blanket the ground smothering crops and causing roofs to collapse. The viscosity of the lava varies and so does its speed, with a maximum of almost 50 km/hr being recorded. The more viscous types of magma (usually acidic) can occasionally prove even more dangerous if their viscosity causes pressure to build up inside the volcano, producing a larger eruption and perhaps a nuée ardente (page 30), the most famous of which occurred on the Island of Martinique in 1902. Here Mount Pelée exploded and clouds of the explosive mixture of gas and tephra, at a temperature of over 800°C, flowed down on the town of St Pierre at 160 km/hr killing all but two of its 28 000 people. Even without this explosive mixture the gas released can cause suffocation. The fine material may become soaked by rain or by water from a crater, and form disastrous mudflows which can flow down onto populated areas (in the eruption of Mount Kelut, Japan in 1919, 5500 people died in this way). The mudflows and tephra can change the course of rivers, block harbours, fill reservoirs and kill fish, as well as destroy bridges. The jökulhlaups of Iceland have already been mentioned, and **tsunamis** at sea (the waves produced by volcanoes and earthquakes) have been known to reach 15 m in height, and because of their momentum they can double that height on reaching the shore. More than

36 000 people are thought to have died from the tsunami from Krakatoa in 1883.

Despite the dangers, volcanoes do have their useful side. If the climate is suitable, many lavas and tephras weather to a fertile soil. This is best demonstrated in some densely-populated parts of the world such as Indonesia, where people are prepared to risk the eruption of a dormant volcano to grow crops on their sides. Elsewhere it has been shown that thin dustings of fine tephra can improve crop yields after an eruption. We have seen that mineral deposits are being formed in the sea at constructive plate margins and diamonds were formed in ancient volcanoes. In addition there are benefits of heat and power to be derived in volcanic regions, and in some cases the areas gain significant income from tourism.

> **Q59.** *Can it be said that volcanoes do more good than harm? (Far more people are killed on the roads!).*

Earthquake hazards

Earthquakes have made a major contribution to our knowledge of the earth's interior (Section 3), but apart from this, and unlike volcanoes, the effects of earthquakes would seem to be entirely negative! Loss of life has been far greater than from volcanic eruptions, and only floods and hurricanes are thought to be more dangerous natural phenomena.

Some knock-on effects are also important. Tsunamis are produced at sea, and on slopes the shocks can trigger off catastrophic landslides and rock falls. In 1970 the ice cap on Mount Huascaran in Peru was dislodged by an earthquake. The ice then swept down the hill picking up rock and becoming a giant debris flow which destroyed the town of Yungay, killing 25 000 people.

In areas of loose sediments the ground may be shaken and compacted causing subsidence. Elsewhere, if fine sediments are saturated, the vibrations of an earthquake can cause the ground to shake the water and rock into a mud (a process called liquefaction). Houses may collapse into the mud, and the sediments may flow or slide down slopes.

Many deaths are caused from indirect effects, such as fire, the collapse of buildings and bridges and the bursting of pipes and reservoirs. Some of the implications of these effects are examined in the next section.

DECISION MAKING EXERCISE

Dealing with earthquakes

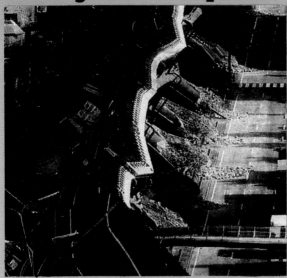

Figure 10.1 *A portion of the Hanshin Expressway collapses after the Kobe earthquake, January 1995*

Earthquake shocks

The intensity of an earthquake is subjectively described with reference to the modified **Mercalli Scale** (Figure 10.2), which is rather like the Beaufort Scale in that it is based on observations rather than upon measurements. It is the modified scale because it was revised in 1931 to take account of the effects on modern urban areas. Should it be modified again?

The greater the energy released in an earthquake, the greater its magnitude. This is measured on the **Richter Scale**, in which the amount of energy released in an earthquake shock is calculated. The scale is logarithmic and open-ended: so far the largest recorded magnitude has been 8.9 (Chile 1960). Estimated comparisons with the modified Mercalli Scale suggest the rough approximations given in Figure 10.3.

MODIFIED MERCALLI SCALE		RICHTER SCALE	EARTHQUAKES PER YEAR
<III	≈	<3.9	830 000
IV to V	≈	4–4.9	62 000
VI–VII	≈	5–5.9	500
VIII–IX	≈	6–6.9	100
X–XI	≈	7–7.9	20
>XII	≈	>8	>0.5

Figure 10.3 *The scales compared*

SCALE	DESCRIPTION
I	Not felt, except by a very few under especially favourable circumstances.
II	Felt only by a few persons at rest, especially on upper floors of buildings. Delicately suspended objects may swing.
III	Felt quite noticeably indoors, especially on upper floors of buildings but many people do not recognise it as an earthquake. Standing motor cars may rock slightly. Vibration like passing of truck.
IV	During the day felt indoors by many, outdoors by a few. At night some awakened. Dishes, windows, doors disturbed; walls make cracking sound. Sensation like heavy truck striking building. Standing motor cars rocked noticeably.
V	Felt by nearly everyone, many awakened. Some dishes, windows etc broken; a few instances of cracked plaster; unstable objects overturned. Disturbances of trees, poles and other small objects sometimes noticed. Pendulum clocks may stop.
VI	Felt by all; many frightened and run outdoors. Some heavy furniture moved; a few instances of fallen plaster or damaged chimneys. Damage slight.
VII	Everybody runs outdoors. Damage negligible in buildings of good design and construction; slight to moderate in well-built ordinary structures; considerable in poorly built or badly designed structures; some chimneys broken. Noticed by persons driving motor cars.
VIII	Damage slight in specially designed structures; considerable in ordinary substantial buildings, with partial collapse great in poorly built structures. Panel walls thrown out of frame structures. Fall of chimneys, factory stacks, columns, monuments, walls. Heavy furniture overturned. Sand and mud ejected in small amounts. Changes in well water. Persons driving motor cars disturbed.
IX	Damage considerable in specially designed structures; well-designed frame structures thrown out of plumb; great in substantial buildings with partial collapse. Buildings shifted off foundations. Ground cracked. Underground pipes broken.
X	Some well-built wooden structures destroyed; most masonry and frame structures destroyed. Foundations cracked. Rails bent. Landslides from river banks and steep slopes. Shifted sand and mud. Water splashed (slopped) over banks.
XI	Few, if any masonry structures remain standing. Bridges destroyed. Broad fissures in ground. Underground pipelines out of service. Earth slumps and land slips in soft ground. Rails bent greatly.
XII	Damage total. Practically all works of construction are damaged greatly or destroyed. Waves seen on ground surface. Lines of sight and level are distorted. Objects are thrown upward into the air.

Figure 10.2 *The modified Mercalli Scale*

The San Francisco earthquake of 1906

This earthquake, which all but destroyed San Francisco, is estimated to have been about 8.3 on the Richter Scale. The San Andreas fault is thought to have moved nearly 7 m, 30 schools and the homes of 250 000 people were destroyed and about 450 lives were lost. Figure 10.4 contains extracts from a contemporary account by a famous novelist and journalist.

"San Francisco is gone! Nothing remains of it but memories and a fringe of houses on the outskirts."

"On Wednesday morning at a quarter past five came the earthquake. A minute later the flames were leaping upward. In a dozen different quarters south of Market Street, in the working-class ghetto, and in the factories, fires started. There was no opposing the flames. There was no organisation, no communication. All the cunning adjustments of a twentieth-century city had been smashed by the earthquake. The streets were humped into ridges and depressions and piles with debris of fallen walls. The steel rails were twisted into perpendicular and horizontal angles. The telephone and telegraph systems were disrupted. And the great water mains had burst. All the shrewd contrivances and safeguards of man had been thrown out of gear by thirty seconds' twitching of the earth crust."

"By Wednesday afternoon, inside of twelve hours, half the heart of the city was gone. At that time I watched the vast conflagration from out on the bay. It was dead calm. Not a flicker of wind stirred. Yet from every side the wind was pouring in upon the city. East, west, north and south, strong winds were blowing upon the doomed city. The heated air rising made an enormous suck. Thus did the fire itself build its own colossal chimney through the atmosphere. Day and night this dead calm continued, and yet, near to the flames, the wind was often half a gale, so mighty was the suck."

Jack London, *Collier's Weekly*, 5 May 1906
from *The Faber Book of Reportage*, 1989

Figure 10.4 *The San Francisco Earthquake of 1906*

The Los Angeles earthquake of 1994

Figure 10.6 is an extract from a newspaper report following the Northridge earthquake near Los Angeles, on 17 January 1994. It is thought that it had a magnitude of 6.8 on the Richter Scale; 61 people were killed and the damage came to $10 billion. This earthquake was not on the main San Andreas fault, but on a smaller thrust fault north west of downtown Los Angeles. The 'Big One' referred to in Figure 10.6 is the major earthquake (perhaps magnitude 8) on the San Andreas fault which is expected soon.

CATEGORY	EXAMPLE
Changing life style	Reliance on electricity distribution grids for pumping water, cooking, communications.
Modern industry	Nuclear reactors. High-tech industry.
Water supply	Long-distance pipes and aqueducts. Dams and reservoirs.
Buildings	High concrete buildings built to sway. Thousands of people in one building. Thousands of glass windows.
Building contents	May contain heavy, tall, metal machines, shelves, furniture etc.

Figure 10.5 *Some twentieth century changes in earthquake risks*

"For the first few nervous days after the quake it did seem as if all was over for the City of Angels. 20,000 people had fled their homes to camp in cars and tents in the city's parks, many because the thought of living between four walls was too terrifying to contemplate. As aftershock after aftershock (several more than 5.0 in magnitude) rippled across the Los Angeles basin from the San Bernadino Mountains to the Pacific's edge, a sense of deep insecurity spread across the city.

"The quake closed down the freeway system in four places, producing three-hour traffic jams and boosting ridership on the commuter Metrolink trains from a measly 1,000 to 25,000 overnight.

"... more than 1,600 buildings bore an inspector's red sticker declaring that they were unsafe, and many thousands more ... were 'yellow tagged', meaning that a person enters at their own risk. Scores of schools were closed by damage.

"Why, for instance, did 12 of the city's hospitals have to close or severely curtail their operations because of quake damage? Why was neither the Santa Monica freeway, nor the giant Golden State freeway, strong enough to withstand a 6.6 earthquake? Why was a major plan to modify the freeways to help them withstand quakes incomplete? Californians know that they need answers soon – before the "Big One" comes along.

Phil Reeves, *The Independent on Sunday*,
6 February 1994

Figure 10.6 *The Los Angeles Earthquake of 1994*

Planning for earthquakes

Planning can take four forms:

● **Prediction**

We are now able to produce reasonably accurate weather forecasts but research into earthquake prediction still has a long way to go. Various techniques have been examined and all have some justification. They range from the behaviour of animals just before an earthquake to measurements of strain and stress along fault lines, the tilt of the ground and the study of electromagnetic waves and water pressure in wells. The principle is that pressure builds up along faults as the earth tries to move, but is retarded by friction. Eventually the force is so great that resistance is suddenly overcome, creating a movement along a fault. For this reason knowledge of the history of earthquakes along an active fault will be important: the longer the area has been without an earthquake, the more likely it is and the bigger the shock will be.

● **Prevention**

Even less progress has been made with research into prevention. The most likely possibility is that by feeding water into active faults it may be possible to reduce the friction and encourage slow steady movement rather than the sudden jerks. Some encouraging results have been obtained from the study of areas where natural lubrication has occurred (e.g. where water has leaked down faults from reservoirs) but the overall results are still inconclusive.

● **Avoidance**

The high cost of earthquake damage as well as the harm to people has encouraged the authorities to draw up regulations to try to limit possible damage. This has involved:
- evaluation of the particular risks in that area;
- producing a model earthquake for the area, based on maximum expected intensity;
- mapping hazard zones for planning and risk assessment purposes.

From this exercise can come regulations about the permitted location and architectural design of buildings, roads, dams etc, as well as plans for dealing with the predicted earthquake and for training everybody concerned in what to do when the expected earthquake arrives.

● **Disaster management**

Most probable sites for earthquakes are now known so that if the will and resources are available a properly prepared plan can swing into action. However not all countries and

The Alaskan earthquake at Valdez

Valdez in Alaska is a small town at the southern end of the 1000 km Trans-Alaska Pipeline where the crude oil is transferred to tankers for the journey to refineries on the west coast of the USA and elsewhere. In 1964 the town suffered damage from the strong Alaskan earthquake even though it was 70 km from the **epicentre**. This was because of the magnitude of the earthquake (about 8.5 on the Richter Scale). Valdez had been built close to sea level on the alluvium of a small delta. When the shock waves arrived, a submarine landslip carried most of the harbour down into the sea, the fine sediments liquefied, and a tsunami, 10 m high travelling at 100 km/hr, carried boats and trees and debris 0.5 km inland. Across the bay oil storage tanks ruptured and caught fire and burning oil was swept out to sea and then back on the waves. Loss of life was limited: it was holiday time (Good Friday evening) and few people were at work or in school; it was a small town in a lightly populated region and most buildings were small and made of wood whilst larger ones were earthquake resistant; and perhaps most important of all, the tide was out at the time of the earthquake. Loss of life at Valdez was small, though 115 people died in southern Alaska as a whole.

Figure 10.7 *The 1964 Alaskan earthquake at Valdez*

regions are well prepared and management of such a disaster is more difficult than it might seem at first.

Use all the information in this section to help you answer the questions as fully as possible.

1 Since 1906 changes made by human beings have made the hazards of earthquakes far greater. Make a list of as many of these changes as you can think of. Study the modified Mercalli Scale and Figures 10.4, 10.5 and 10.6 will give you further ideas.

2 Perhaps there are advantages in being unable to produce accurate predictions. How should the USA President react to a firm prediction that in a month a major earthquake would strike San Francisco? What would the practical problems be if he announced that prediction?

3 After the Valdez earthquake it was decided to rebuild the town on a new site nearby. Write a specification for the sort of site required.

4 Is it reasonable to plan for the permanent and wholesale evacuation of people and businesses from areas with a known high earthquake risk?

5 Imagine that your home area was one likely to experience a major earthquake. Draft a plan for establishing a control and co-ordination centre; outline the job descriptions for members of the team; and set out major provisions to cope with the earthquake when it comes.

- A large earthquake shock destroys most of the usual communications. Power supplies and lines will be lost, roads and possibly airstrips damaged, and so vital information will be hard to obtain.
- In the cities, such roads and streets which have not been destroyed may be jammed with traffic.
- Water supplies may have been seriously damaged with implications for the control of fire and disease.
- Some hospitals may not have survived and those which have will also have to assemble staff.
- The first victims brought in may not be as badly injured than ones who arrive later, forcing difficult decisions about who should be sent away.
- The bodies of victims have to be dealt with and this is not easy if laws require that they be identified and certified dead by doctors who should be treating the living.
- Supplies of aid will soon start to arrive, placing further strains on the administrators. Some of the aid will be welcome but much is likely to be inappropriate and carrying political burdens.
- The airports and warehouses and struggling transport systems become clogged with unnecessary materials while large numbers of people have to be organised (and fed and paid) to deal with all this.
- The disaster managers will be exhausted and need to be relieved for at least several hours of sleep.
- There will be the longer-term problems of food and shelter and rebuilding the shattered economy.

Figure 10.8 *Problems of disaster management*

PROJECT SUGGESTIONS

- For the next (or more recent) earthquake or volcanic disaster, collect a series of reports and analyse them after a few weeks. Relate the causes to the plate tectonic situation of the area, and note especially how the perception of deaths and destruction changes as more accurate information becomes available.

    To what extent were the citizens prepared **a)** in terms of resistant buildings, communications etc.; **b)** for the post-earthquake problems which had to be faced?

    Assess the validity of any criticisms which were made.
- For an area known to you, obtain the geological survey maps (from a local library or museum) and relate the rock types to the shape of the land.

You may find it helpful to trace some geological outcrops onto tracing paper and lay this over the contours of the Ordnance Survey maps.
- Carry out a similar exercise for an area of coastline including cliffs and lowlands noting how geological structures (folds and faults) as well as rock types may affect the coast. Remember though that cliffs are not necessarily made of hard rock (the chalk cliffs are soft) but are merely hills at the coast and they may have other reasons for being there.
- From standard geology textbooks (References below) find out more about the geological aspects of fuel supplies, mineral deposits; and engineering projects such as the construction of roads and dams, and of the Channel Tunnel.

REFERENCES

A recent and very readable account of the development of the British Isles is:

Fortey, Richard (1993), *Hidden Landscape*, Jonathan Cape

An essential reference book for A level geographers, with excellent diagrams and clear definitions is:

Whittow, J (1984), *Dictionary of Physical Geography*, Penguin

The most useful dictionary of Geology is:

Lapidus, DF (1990), *Dictionary of Geology*, Collins

A simple introduction to GCSE level geology is:

Webster, D (1987), *Understanding Geology*, Oliver and Boyd

A more advanced text is:

Bradshaw, M (1980), *The Earth: Past, Present and Future*, Hodder and Stoughton

The standard A level Geology text is:

McLeish, A (1986), *Geological Science*, Blackie

The following large volumes should be in a good library; the first is particularly recommended for its up to date text and for its excellent coloured diagrams:

Skinner, BJ and Porter, SC (1987), *Physical Geology*, Wiley

Duff, D (Ed) (1993), *Holmes' Principles of Physical Geology*, Chapman and Hall

Smith, DG (Ed) (1982), *The Cambridge Encyclopedia of Earth Sciences*, CUP

Press, F and Siever, R (1986), *Earth*, Freeman

USEFUL ADDRESSES

The Geology Section of the Natural History Museum has excellent modern exhibits relevant to most of this text, and its shop has a wide range of guide books, maps and specimens.

The Natural History Museum Earth Galleries, Exhibition Road, London SW7 2DE.

Local Museums are worth seeking out (ask at the local library) as they often have good local collections and exhibits.

The British Geological Survey sells maps and publications covering the whole country.

The British Geological Survey, Keyworth, Nottingham, NG12 5GG.

The Geologists' Association is the main body for the amateur geologist. It arranges field trips and publishes its own journal and guide books.

Burlington House, Piccadilly, London WIV 9AG.

Epidemiology and Management of Rice Tungro Disease

The Outcome of the Rice Tungro Disease Management Workshop
11–14 November 1993, Alor Setar, Malaysia

Edited by
T.C.B. Chancellor and J.M. Thresh

WITHDRAWN
Formerly
UNIVERSITY
AT
MEDWAY
LIBRARY

UNIVERSITY OF GREENWICH LIBRARY
633.
1898
RIC

© *The University of Greenwich* (1997)

The Natural Resources Institute (NRI) is a scientific institute within the University of Greenwich, and is an internationally recognized centre of expertise in research and consultancy in the environment and resources sector. Its principal aim is to increase the productivity of renewable natural resources in developing countries in a sustainable way by promoting development through science.

Short extracts of material from this publication may be reproduced in any non-advertising, non-profit-making context provided that the source is acknowledged as follows:

Chancellor, T.C.B. and Thresh, J.M. (eds) (1997) *Epidemiology and Management of Rice Tungro Disease*. Chatham, UK: Natural Resources Institute.

Permission for commercial reproduction should, however, be sought from the Head, Publishing and Publicity Services, Natural Resources Institute, Central Avenue, Chatham Maritime, Kent ME4 4TB, United Kingdom.

T.C.B. Chancellor currently is an Affiliate Scientist at the International Rice Research Institute, PO Box 933, 1099 Manila, Philippines

This publication is funded by the Department for International Development (formerly the Overseas Development Administration) of the United Kingdom. However the Department for International Development can accept no responsibility for any information provided or views expressed.

Price £10.00

No charge is made for single copies of this publication sent to governmental and educational establishments, research institutions and non-profit-making organisations working in countries eligible for British Government aid. Free copies cannot normally be addressed to individuals by name, but only under their official titles. When ordering please quote **PSTC27**.

Natural Resources Institute
ISBN: 0-85954 433-8
ISSN: 0952 8245